the Natural Environment: Wastes and Control

the Natural Environment: Wastes and Control

R. F. CHRISTMAN
B. W. MAR, E. B. WELCH
R. J. CHARLSON, D. A. CARLSON

Department of Civil Engineering
University of Washington
Seattle, Washington

Goodyear Publishing Company, Inc.
Pacific Palisades, California

Current printing (last digit):

10 9 8 7 6 5 4 3 2 1

ISBN 0-87620-610-0

Library of Congress Catalog Card Number 72-97277

Y-6100-5

Printed in the United States of America

To Professor Robert O. Sylvester,
a colleague and friend, whose
leadership and pioneering efforts
in interdisciplinary education
have inspired each of us.

Contents

PREFACE

Many college and university students are deeply concerned over the environmental problems mankind faces. This concern first grew among students of the various professional programs and spread to students of all subjects who accurately sensed the emergence of problems unparalleled in human history and which transcend all traditional disciplinary boundaries. The rising concern is even beginning to gnaw on the national conscience, although the evangelical zeal of the first advocates of environmental reform has subsided. In its wake lies not only increased awareness, but confusion regarding the seriousness of the problem. Are the citizens of the United States prepared to spend nearly $300 billion over the next decade to cleanse their environment? Is this amount adequate? Is it necessary?

Since 1969, we have had the privilege of offering an undergraduate course at the University of Washington entitled, "Man and the Pollution of His Environment." The course has been open to all students of the University, graduate and undergraduate alike, and has had no prerequisites. We have been deeply impressed with the ability of the students taking this class to put aside environmental breast-beating and to search rationally for a fundamental understanding of the natural environmental mechanisms upon which human activity impinges. They have expected clear descriptions of the nature of human waste products and their known effects on the environment. They have learned what alternatives technology offers in waste management methodology and what the cost of each alternative is. Above all, they have been anxious to distinguish between established facts and speculation.

This text has evolved from our experience with this course. It is designed to present factually what is known regarding water pollution, air pollution, and solid waste management. In addition, it attempts to describe the most serious problems this nation faces in deliberating future policies affecting its air, water, and land resources.

Russell F. Christman
Brian W. Mar
Eugene B. Welch
Robert J. Charlson
Dale A. Carlson

1 The Natural Environment

ORIGIN OF OUR PLANET

The reader may wonder, with justification, why a text concerned with the mechanics of environmental pollution should begin with a section on the origins of the planet earth. The reason, simply, is that the random chemical and physical reactions which scientists believe led to the formation of the planet billions of years ago are still active today and actually constitute the vital machinery which much of human activity threatens to alter. Thus the often subtle environmental effects of pollution cannot be separated from the natural geophysical activities which originated and maintain our natural environment. To gain a proper perspective on the effects of human activity on "nature," one must first understand the mechanics of an environment that was formed and shaped in the absence of man by a variety of complex natural forces.

Certainly, all ages of men have felt obliged to explain the formation of the earth. At first, their explanations were mythical and often dogmatic, covering the details of observable nature from the stars to the fishes of the sea. As man's understanding of the material universe grew, the scientific basis for what he believed of the earth's formation broadened. Man began to discover other planets, explain the tides, transform chemicals into new substances in laboratories, and understand natural transformations outside of laboratories. Alternative explanations of the earth's formation became possible. We should note, however, that both religious and scientific views are inadequate explanations, since we do not have an eyewitness account and many questions can still be raised.

What might have happened? Can science with its current understanding of physical processes offer a model of the earth's formation, one which is consistent with known physical laws and which explains how things might have come to be the way we see them today?

The answer is yes, if scientists begin with a few fundamental substances called protons, neutrons and electrons and explain how these reacted with each other, and how the products of the reactions could have led to the formation of our planet. Scientists do not say it did happen this way, only that it might have, and that, if it did, no known scientific law would have been violated. A simplified but complete account of this theory follows.

Through complex nuclear reactions, all of the elements were formed, the lighter elements—those with the smallest atomic weights, like hydrogen and carbon—in the greatest abundance. The elements began to accumulate into dust-like particles, and under the influence of small gravitational forces pre-planets took shape. Many of these "planetesimals" drifted together to form the earth. As the earth dust-ball grew in size, gravitational pressure toward the interior increased greatly, resulting in very high temperatures as compaction continued. High pressure and temperature produced even more complex nuclear reactions, which in turn raised the temperature still more. The early earth was probably never completely molten, nor did it initially possess a substantial atmosphere of its own. The nuclear reactions going on in the interior caused the lighter elements (C, N, O, H, S, P, etc.) to float toward the surface while the heavier elements sank toward the center. When these light elements reached the earth's crust, they burst through in volcanic eruptions, bringing some solid material with them. The lighter gases like hydrogen probably escaped into space, leaving the heavier gases, like nitrogen, in the early atmosphere. Scientists estimate that these events, that is, the reactions and accumulations that led to the earth's formation, began around 5 billion years ago.

The inadequacies of the scientific view become very clear at this point. It is uncertain, for instance, whether the newly formed earth was covered with a very dense atmosphere containing large quantities of water, carbon dioxide, nitrogen, and other components, or whether the atmosphere that we know today resulted only from very slow leakage of gas from the interior of the earth for billions of years. The latter theory is currently more popular among scientists, since an early dense atmosphere would probably have left a dramatic record in the earth's geology which has not been observed.

Two elements, hydrogen and oxygen, probably existed prior to the dust accumulation stage. These elements react spontaneously to form water, especially at the high temperatures that prevailed at that time. Water probably existed exclusively as a vapor in the hot regions of the early earth, and as the overall cooling process continued, it condensed to form our first bodies of water. The reaction of oxygen with hydrogen, as well as other reactions with materials in rocks, kept our atmosphere from building up appreciable concentrations of gaseous oxygen. As the millions of years passed, one can imagine a more or less continuous process of volcanic emission and subsequent condensation of water steadily increasing the volume of water contained in oceans and lakes. Water present at this time was not pristine pure, since it interacted with the gases in the atmosphere and with the rocks on the surface of the earth. In fact, it probably contained dissolved materials in compositions very much like what we see today.

Actually the movement of dissolved materials through our planet's oceans is a continuous sedimentation of minerals from the water to the ocean floor and a re-solution of minerals to take the place of what was precipitated. Scientists have observed this phenomenon, and their measurements indicate that the sedimentation rates are in fact much greater than the re-solution rates that can be predicted. This would mean, of course, that the earth's oceans would in time lose their mineral content, if it were not for the fact that volcanic eruptions and emissions of hot springs continue to contribute dissolved material to the oceans today as they have throughout the history of our planet. It is important to realize that although the amounts of air in our atmosphere and water in our oceans gradually increased during the first few billion years of the planet's existence, the quality of these environmental resources has remained remarkably constant.

During these early times, intense ultraviolet radiation showered down on the surface of the earth. Radiation of this magnitude fractures the water molecule into its component gases, hydrogen and oxygen. This process represents the first source of free oxygen gas in our atmosphere, although probably not the most important one since it is a self-limiting process. That is, oxygen (as O_2 and O_3) itself absorbs ultraviolet radiation, and as soon as its concentration increased slightly in our atmosphere, insufficient radiation reached the surface to fracture any more water molecules. Scientists refer to this process as photodissociation and they do not believe that it continued once the concentration of oxygen became greater than one percent of its present concentration. It is because water absorbs ultraviolet radiation that scientists think the first life forms on this planet originated in the sea.

Photosynthesis is a process which produces large amounts of oxygen by reacting light, carbon dioxide, and water, all of which were present in that early period. The process of photosynthesis could have occurred only if life existed at that time, bringing up the strong suggestion that life-producing metabolic events began at about the end of the period of photo-chemical oxygen production. Geochemists date this period at between 2 and 3 billion years ago. The first life, whatever form it was, possibly primitive bacteria, must have occurred under water in shallow ponds where the ultraviolet radiation could not damage it. At a later time, photosynthetic organisms appeared. As soon as the process of photosynthesis began to build up the oxygen concentration, the oxygen itself began to absorb ultraviolet radiation, producing ozone in the upper atmosphere. Thus the lethal radiation was greatly attenuated, permitting a diversity of life forms at the earth's surface and removing the need for life to exist only under water.

One inescapable fact confounds this picture: simple life forms—even our yeasts of today—find oxygen toxic. It is possible that the first life forms were anaerobic like yeasts and bacteria, and the first photosynthesis may have produced something other than molecular oxygen. As evolution proceeded to a point where appropriate enzymes existed for protecting simple life forms from toxic oxygen, the gas could then be produced in quantity by other organisms. It seems clear to geochemists today that photosynthesis is *currently* the major source of atmospheric oxygen, and that it is probably responsible for a substantial part of the build-up of this gas over the last few billion years. Just how much oxygen came from photodissociation and how much came from photosynthesis is not known with any precision.

What we do know is that today oxygen is recycled through the biosphere about once every 2,000 years by the processes of photosynthesis, metabolism, and oxidative decay of dead plants and animals. We also know that the part of the earth's oxygen that has been cycled through the biosphere almost exactly equals the amounts of carbon in sediments (as coal, oil, and carbonate rocks). This balance, coupled with the intimate linking of plant life to the presence of O_2 in air (on which mammalian life depends), emphasizes the indivisibility of the biosphere. See Figure 1–1.

The land masses of the early planet were very different from today's islands and continents, and the amount of volcanic activity was certainly much greater. Continents were not eroded to the present extent, and they were probably in different locations and of different shapes. In fact, scientists believe that they are still changing. We must realize that in these early periods weather did exist, although it was perhaps very different from what we know today. Winds blew, rain and other precipitation occurred, and early in the earth's history, perhaps millions of years ago, ice ages occurred. Both water and ice action grind rocks and other crustal materials into powder, so that particles can be moved by rivers in the form of sediments; the soluble parts of the minerals are eroded to a much greater extent than the insoluble parts and thus provide nutrient materials for life. The drifting and tilting of the continents caused land to submerge and re-emerge from the oceans, bringing all manner of primoridal life forms into new situations where they could evolve. A combination of all of these processes over geologic eras made the chemical composition of our water and air systems suitable for the creation and sustenance of life on this planet.

Several sensitive chemical and physical phenomena maintain the life-sustaining conditions in our atmosphere and oceans. One of these involves the extremely temperature-sensitive vapor pressure of water. We can imagine that if the earth were much warmer or much cooler, our weather and all things dependent on it would be greatly different. Another important regulatory phenomenon is the equilibrium between the amount of CO_2 gas in the

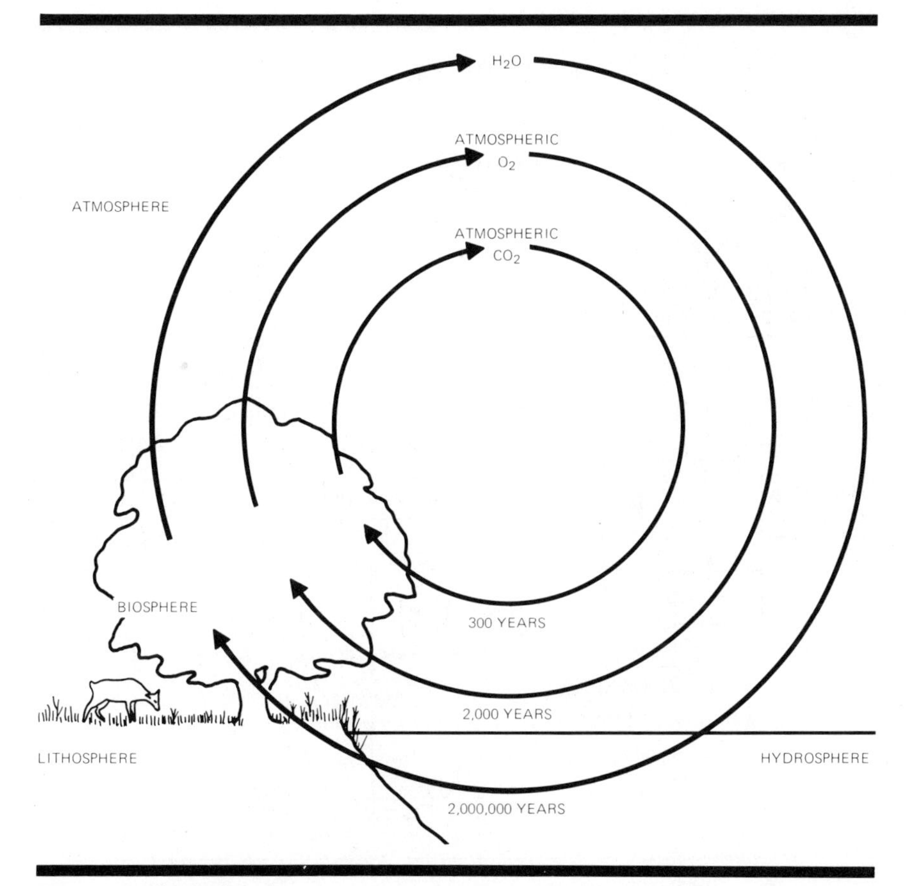

Figure 1–1. Chemical Cycles

This diagram shows in simplified form the cycles by which the biosphere, atmosphere, and hydrosphere exchange water vapor, oxygen, and carbon dioxide. The approximate time span of each cycle is also indicated.

atmosphere and in the water systems on the planet as well as in the sediments under those water systems. These equilibria are responsible for the maintenance of a rather delicate level of acidity in the earth's bodies of water.

The temperature of the earth itself is governed by many factors, including the amounts of water and carbon dioxide in the atmosphere, as shown in Figure 1–2. Clouds and dust particles reject some of the visible light coming to the earth from the sun, our only source of incoming energy. Clouds absorb outgoing infrared light, as do water vapor and carbon dioxide. Thermal balance occurs when the incoming radiation and the outgoing radiation are equal. Variations in the amounts of any of these substances affect the temperature of the earth.

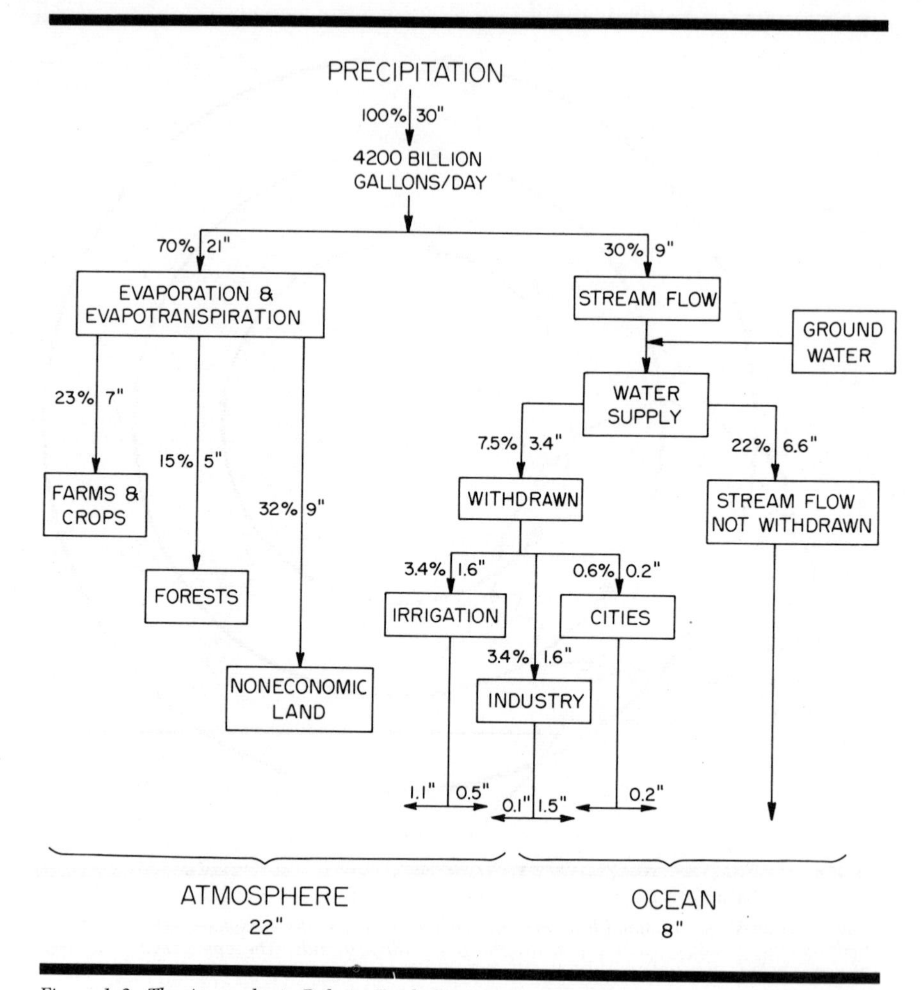

Figure 1–2. The Atmospheric Role in Earth Temperature Regulation

Evidence indicates that man can substantially modify his environment, perhaps modify it to a point where life itself would not be possible. It has been suggested that by the addition of substances to the atmosphere man is unwittingly conducting a vast geophysical experiment. One could only speculate as to what the consequences would be if life disappeared from the earth. Would our planet become cold and dry like Mars, or hot and humid like Venus?

Somewhat contrary to the usual notion of life evolving because of stable environmental conditions, we have painted a picture of the impact of the evolution of life itself on the formation and characteristics of the environment. In later sections, we will focus our discussion on the effects of life, and in particular man's many activities, on the maintenance of suitable environmental conditions.

THE ATMOSPHERE AS AN ELEMENT OF THE ENVIRONMENT

Man's involvement with the earth's present atmosphere extends far beyond his dependence on certain chemical components of the air. The atmosphere shields man from lethal cosmic and ultraviolet radiation and burns up showering meteorites. Atmospheric winds that transport heat and moisture tend to create more uniform conditions on the face of the earth than would otherwise exist. These same winds drive the ocean currents, erode the soil and transport pollen. Most of the sounds man hears, the sights he sees, and the odors he smells are inextricably connected to the state of the atmosphere. Man tends to think of the atmosphere as a limitless reservoir, but this assumption is more wishful thinking than sound scientific fact: the volume of the atmosphere is not only limited but is surprisingly small.

This volume limitation occurs primarily for two reasons. First, the total atmosphere is a relatively small body of gas; if all of the atmosphere were brought down close to the earth so that a uniform pressure of 760mm Hg (one atm.) were maintained at all points, the atmosphere would extend approximately 5 miles from the earth's surface to the top. Second, gases and particles in the atmosphere are distributed slowly and never evenly mixed. For example, the stratosphere still contains the debris from atomic bomb testing of years ago in concentrations beyond what would have existed if the atmosphere were well mixed. Other and perhaps more familiar examples are the air pollution crises common in such metropolitan areas as Los Angeles, London, and New York. In these cases the lack of mixing, which decreases effective volume, is caused by warm air lying on top of cooler air. Meteorologists call this condition a temperature inversion, the reverse of what ordinarily occurs—temperature decreasing with height above the ground. Because of this effect, mixing may extend only a few hundred feet above the surface. We are all too familiar with the tragic consequences of the heavy introduction of pollutants to the atmosphere under these meteorological conditions. The fact that these effects may be only local in nature and of short duration does not lessen the impact on those living under it.

The blanket of air surrounding the earth today is a complex mixture not only of gases but of liquid and solid material. To the scientist, the complexity of this mixture presents extremely severe analytical problems, because the constant motion of the gas caused by temperature-gravitational effects makes measurements difficult. Nonetheless, it has been established that the fixed gas composition of the atmosphere is remarkably constant throughout its entire volume. This can be seen in Table 1–1.

Minor constituents of the atmosphere often vary in concentration. Water may vary from 5 percent to less than 0.1 percent and is the only atmospheric component that exists in all three phases—solid, liquid, and gas. The cycle of evaporation and precipitation is an effective way of transporting heat over great distances, in that energy is taken up during evaporation and released during condensation. Carbon dioxide is sometimes considered a variable constituent and is produced in greater quantities over cities. There is

TABLE 1–1

Fixed Gas Composition of Earth's Atmosphere

GAS	% BY VOLUME
N_2	78
O_2	20.9
CO_2*	0.03
Ar	0.93

*Sometimes considered variable

some concern in the scientific community that the concentration of CO_2 in the atmosphere is increasing because of man's infatuation with the combustion process, and that this increased concentration may cause slight increases in the surface temperature of the earth. Even slight increases may be enough to change fish migration habits in the oceans and to begin melting the polar ice caps. There is also speculation that a cooling trend will result from the increased concentration of atmospheric particulate matter attributable to man's activities.

The occurrence of ozone in our atmosphere has already been mentioned. It is found in very minute quantities near the surface of the earth, and the concentration increases with altitude, reaching its maximum at a height of approximately 20 miles. This gas is produced by the interaction of intense solar radiation, oxygen molecules, and other molecular species. It is not commonly realized that the presence of solid particles suspended in the air is a natural occurrence in the atmosphere. Dust particles are swept up by the wind from exposed soil, fires, and volcanos. In many cases, particles are created by the injection of salts into the atmosphere when ocean spray evaporates. Large particles settle out of the atmosphere fairly rapidly, but there are many small particles, too small to be seen by the naked eye, that may remain suspended for relatively long periods of time. The number of dust particles in the atmosphere at any one time or place is quite variable. However, it can be said that smoke or dust concentrations become very great over large urban or industrial centers.

The motion of the atmosphere is very complex and in many ways still evades precise analysis by the meteorologist. In general, air does not move from regions of high pressure to low pressure, but spirals or rotates around regions of high and low pressure. The reason for this rotational motion is closely linked to the fact that the earth itself rotates. High and low pressure centers are distributed around the entire earth and are themselves in a state of constant motion. There exist several probable patterns of highs and lows around the earth which cause our familiar regional climatic conditions. Climate in this sense includes the prevailing winds, the range of temperatures, precipitation, humidity, and so on. Much of what we call weather results from the mixing of air from two different sources, for instance, from the arctic and from the tropics. Such mixing is called frontal activity, and the region of mixing itself is called a front. If one air mass is moist and warm and the other cool, fronts frequently cause precipitation in the form of rain or snow. The recent advent of such earth satellites as Nimbus, ATS, and the ESSA series has resulted in useful and interesting photographs of atmospheric motion which will increase our ability to predict weather.

Atmospheric motion is essential to life on earth. It is obvious that winds carry away pollutants and help to disperse the temperature inversions in urban areas. The hydrologic cycle is totally dependent on the movement of water from place to place, and solar energy is distributed by atmospheric motion to the cooler regions of the earth, resulting in a more or less uniform terrestrial temperature. The temperature of the atmosphere itself is a function of altitude as shown in Figure 1–3. It is clear from the earlier discussion of oxygen evolution that the "stuff" of the environment is not static, but is continuously cycled. There are a number of important cycles that need to be mentioned; the water cycle is probably the most obvious and important one. Besides this cycle, essential cycles exist for oxygen, nitrogen, and many trace chemical materials. We will begin with water and proceed to the rest, while realizing that these cycles are in fact interwoven and really quite indivisible.

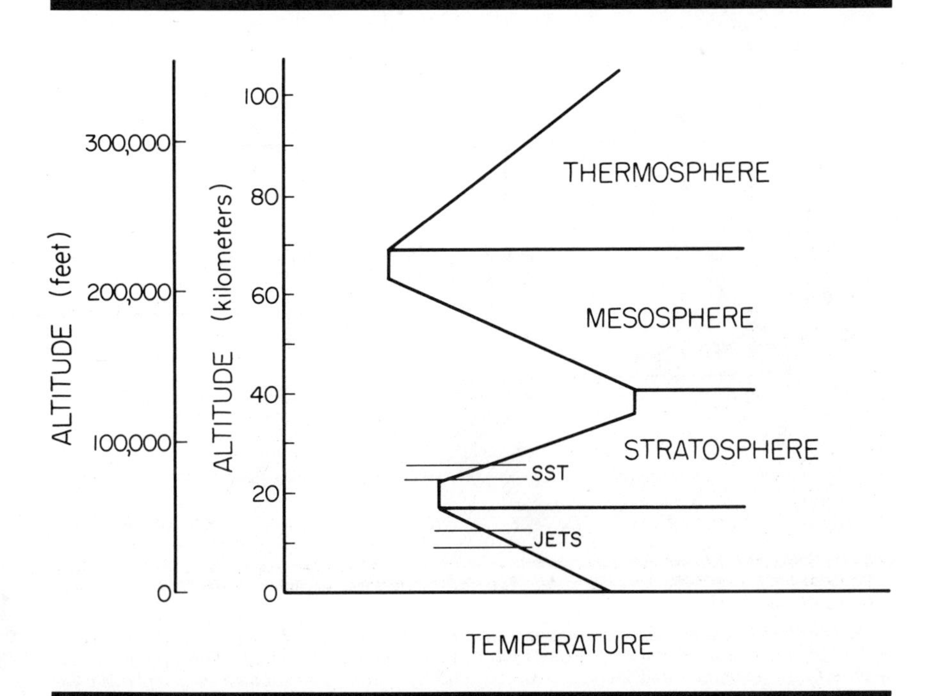

Figure 1–3. Variation of Temperature with Altitude

THE HYDROLOGIC CYCLE

Man's dependence on water environment extends well beyond the water he needs for his bodily functions. Water is needed by man for power, transportation, irrigation, and a multitude of other purposes. More importantly, the water environment must be controlled by man to protect himself from floods, rainstorms, avalanches, and droughts. What we refer to as the natural hydrologic cycle is a collection of processes that control the amount of water in any place at any time. Man's efforts to modify this cycle for his benefit have produced major engineering structures such as dams, bridges, irrigation projects, coastal breakwaters, and reservoirs.

A major feature of the hydrologic cycle is the conservative nature of water; that is, it is not destroyed by use, as are mined minerals or fossil fuels. Therefore water is a renewable resource, and the earth will never go dry as a result of man's use of water. This simple fact is not widely recognized, probably because of our keen awareness that man's use of water can seriously affect its quality.

Basically, the hydrologic cycle involves the vaporization of water from the earth's surface and its subsequent condensation and precipitation over land and water masses, as shown in Figure 1–4. During the process of vapor-

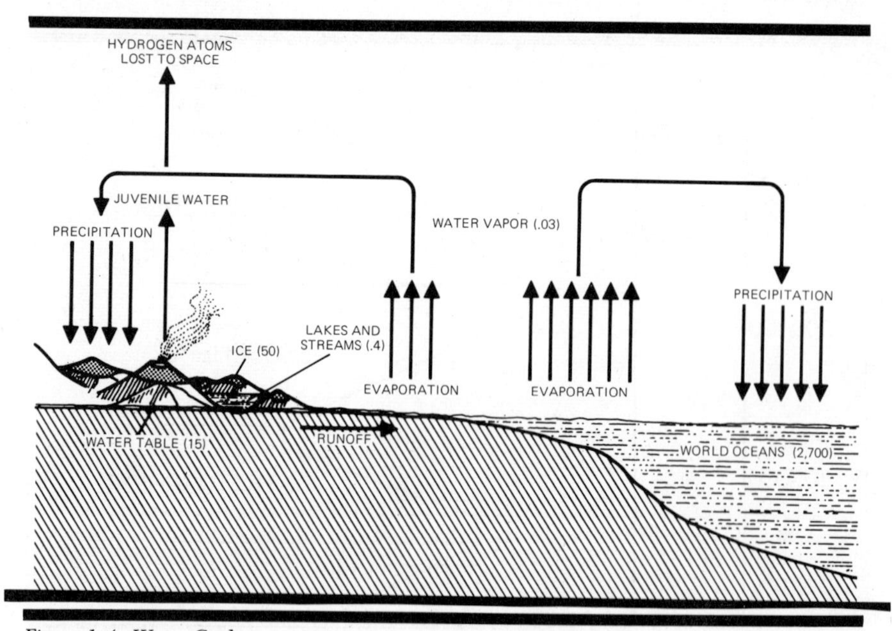

Figure 1–4. Water Cycle

In the biospheric water cycle, evaporation and precipitation must be equal on a worldwide basis. Although ocean evaporation exceeds return precipitation, for example, the opposite is true of terrestrial evaporation and precipitation. On land this excess precipitation may replenish the water table, enter lakes and rivers to return to the sea as runoff, or become ice caps and glaciers (75 percent of all fresh water). Minimum estimates of amount of water in each reserve are indicated as depth in meters per unit area of the earth's surface.

ization, water is purified of any nonvolatile materials such as dissolved salts. Some water precipitates on high ground and returns via streams and underground flow to the ocean. Thus the hydrologic cycle serves as a pump moving large quantities of water from low elevations to high elevations. The flow back to the sea constitutes a tremendous power resource which engineers can tap at various stages by constructing dams. This return flow to the ocean transports sediments and dissolved materials which are accumulated by the erosion of the land. These dissolved materials cause the water to deteriorate gradually in quality during its return to the ocean.

To develop a proper sense of the magnitude of the earth's water resource, it is necessary to examine the relative sizes of some numbers and become acquainted with certain units of measurement which may be unfamiliar. Since we are concerned most closely with the water on our land and its transportation, we might begin by looking at the amount of land we are talking about. As can be seen from the data in Table 1–2, most of our land is drained by river systems which ultimately carry runoff to the oceans. Although such information is interesting, it tells us nothing about the amount of water being drained in any time period. Furthermore, one might imagine that the amount of runoff is somewhat less than the total rainfall, since some water is lost through evaporation before it can be returned to the ocean.

A unit of water quantity used frequently by engineers and scientists is the acre-foot—the volume of water required to cover to a depth of one foot one acre of level ground. One acre-foot is equivalent to about 326,000 gallons. A little over three acre-feet of water would satisfy the water needs of 5,000 to 10,000 people for one day, at the level of typical use in the United States.

How much rain falls to the earth's surface in the course of one year? The best estimate we have is around 380 billion acre-feet. Most of this falls into the earth's oceans (300 billion acre-feet). The 80 billion acre-feet that falls on land is drained by the systems described in Table 1–2. The major and minor river systems together carry about 28 billion acre-feet; only 0.3 billion acre-feet are drained to inland seas and lakes. The remainder, 53 billion acre-feet, or nearly two thirds of the total rainfall on land, evaporates and does not appear in the drainage networks. Although the numbers appear to be almost incomprehensibly large, it is worth emphasizing that only 34

TABLE 1–2

Water Drainage of the Earth

EARTH LAND	BILLIONS OF ACRES
Drained by 68 major rivers	14
Drained by minor rivers	11
Drained by inland seas and lakes	8
Covered by glaciers	4
Total Earth Land	37

percent of land rainfall and only 7 percent of the total earth rainfall is in any way usable by man.

Looking at the water quantity problem in another way, if the total quantity that falls on the United States in an average year were spread evenly over the country, it would stand to a depth of approximately 30 inches. Some of this water returns directly to the atmosphere, some seeps into the ground, and some runs off through stream channels. Approximately 21 inches of this water is returned directly to the atmosphere without entering rivers and streams. The remaining 9 inches is our usable supply and drains from the land surface into rivers and streams. From this 9 inches, approximately 3 inches of water are currently withdrawn for municipal and industrial purposes. Of this 3 inches, 1 inch is consumed or returned to the atmosphere during use, and 2 inches are returned to the streams with decreased quality. This leaves approximately 8 inches of the original 30 flowing into the ocean. See Figure 1–5.

The 9 inches that reach the streams and rivers are referred to as "runoff." If you were to examine the distribution of runoff over the continental United States, you would observe that the average 9 inches does not occur in all regions. There is a tendency to have large quantities of runoff in the southeastern United States, the northeastern United States and in the Pacific Northwest. Average runoff occurs in most of the eastern United States, and much less than average runoff occurs in the southwest and the midwest. Part of the runoff in the streams will, of course, seep through the ground to form an underground water supply. Underground water provides approximately 20 percent of the 3 inches of water drawn nationally for municipal and industrial use. The remaining 80 percent is obtained from lakes and streams directly.

In addition to geographic variations in the amount of rainfall, there are temporal variations. Differences in annual rainfall on a given region can vary from approximately 50 percent to 150 percent of average. These variations result in periods of drought or floods. Other variations are seasonal, i.e., winter rainfalls or summer droughts, and may result in runoff variations as great as a factor of 10. There are additional complications in relating precipitation to runoff; precipitation that falls in the form of snow in the winter months may not melt until later in the year, and would not be recorded as stream flow for as long as 4, 5, or 6 months later. Precipitation on lower elevations will return immediately as runoff and can create very serious problems. An example of this occurred in southern California during the winter of 1969. Excessive amounts of rainfall created flooding, mud slides, erosion, and other serious problems to the local environment.

We have seen that the natural hydrological cycle delivers and distributes water unevenly in time and space. If man elects to live in regions of excessive or deficient rainfall, he must erect engineering structures, often of tremendous magnitude, to normalize these elements of his environment throughout the year. In order to determine the actual magnitude of such engineering structures, we must discuss the various uses and demands that man places on water resources through his various activities.

The major uses of water by man are for irrigation, industrial operations, and what might best be termed domestic use. Of these uses, irrigation and industrial operation are the largest, accounting for approximately 92 percent (46 percent each) of the water used, whereas domestic consumption is on the order of 8 percent. The amount of water consumed, that is, evaporated or actually incorporated into a product, is one of the most important characteristics of water use. Irrigation uses alone account for 80 to 90 percent of all water that is consumed. See Figure 1-5.

The record of water use by regions in the United States reflects the economy of our nation. Agriculture in the West has to depend on adequate water for irrigation, whereas industries have concentrated in the East partly because of the existence of plentiful water supplies. The small return flows from irrigation practices intensify the water problems of the West by depleting the quantity as well as degrading the quality of water supplies. The 3 inches that we mentioned earlier—those that are used for municipal-industrial purposes in the United States—will rise by the year 1980 to almost 7 inches, and increases in the eastern states will be almost four times those of the western states. This is because the western states will probably be using water predominantly for irrigation, whereas the eastern states will be using water for industry, and the rate of industrial development exceeds that of agricultural development. Therefore, the greatest increases should occur in non-consumptive uses of water.

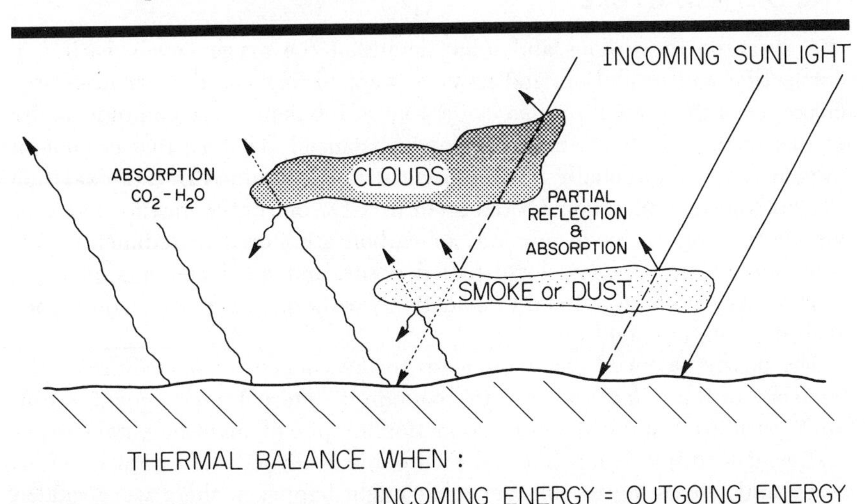

Figure 1–5. Use of Water in the United States

From ENGINEERING MANAGEMENT OF WATER QUALITY, by P. H. McGauhey, p. 76, McGraw-Hill, 1968. Reprinted by permission of the publisher.

The uses of water mentioned so far involve only water which is actually withdrawn from the rivers or streams. In addition, there are certain in-stream uses of the water, which include the production of electricity, the transportation of goods, recreation, and fish and wildlife preservation. Each of these uses makes a specific demand on water quality and water quantity. As you may imagine, the water used by cities and industries is returned containing quantities of waste, which generally degrades the water quality and inhibits the use of water for some in-stream purposes.

Thus a conflict can arise between withdrawal uses of water and certain in-stream uses which demand high water quality for proper development or enjoyment. Major policy decisions are required to equitably allocate the water resource to each need.

Additional conflicts can develop when man attempts to control the hydrologic cycle. A dam to store water during high flow also increases water availability during low flow periods. Thus dams are beneficial, since they can help reduce flood damage, increase the amount of water available for irrigation, and reduce currents in a river. On the other hand, dams can be costly, since they prevent fish passage, inundate lands behind the dam, change environments to favor less desirable kinds of aquatic life, provide large surface areas for evaporation losses, and cause water to stratify and otherwise degrade in quality. Each engineering solution to an environmental problem will have advantages and disadvantages of this kind which must be foreseen and evaluated.

THE OXYGEN CYCLE

We discussed the origins and some details of the oxygen cycle earlier in connection with evolution, and now we want to focus on the "steady state" character of the present oxygen cycle. Figure 1–6 depicts the complex nature of this cycle, which is almost perfectly balanced. That is, the amount of oxygen produced annually by photosynthesis is consumed by the oxidation of plant and animal carbon. Oxidation forms CO_2, one of the starting materials for photosynthesis. Small amounts of carbon are stored in sediments, e.g., peat moss, which become fossil fuel deposits, and small amounts of atmospheric oxygen are consumed in oxidative weathering of rocks, leading once again to a static condition.

The nearly balanced character leads to the concept of *homeostasis*, "the tendency of the living organism to maintain its internal environment within limits permitting survival and reproduction, in spite of environmental stresses that tend to displace the internal environment beyond these limits."[1] It is useful and interesting to consider what might happen if there were sudden changes in the "internal environment" of the oxygen cycle, which is actually a fragment of all organisms. A decrease in the O_2 demand of decaying plants, for instance by removal of large numbers of plants from the earth's surface,

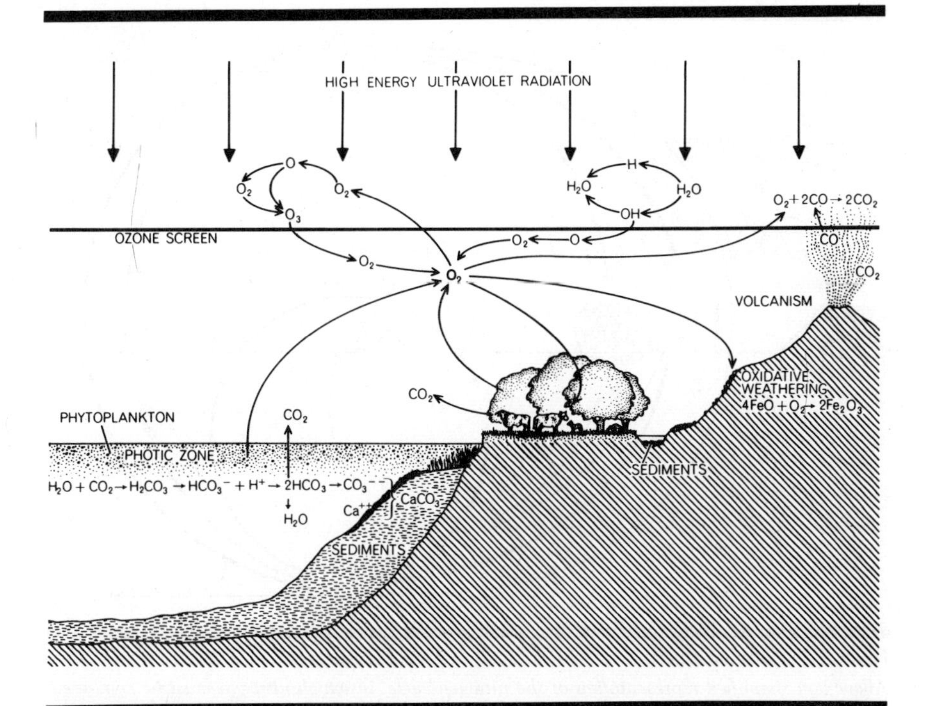

Figure 1–6. Oxygen Cycle
A few of the pathways of the complex oxygen cycle are shown here in simplified form. Among many chemical forms and combinations of oxygen are molecular oxygen (O_2), water, and organic and inorganic compounds.

would result in an increase in O_2. Subsequently there would be a more rapid oxidation of the remaining dead and dying material due to the higher O_2 content and a decreased amount of O_2 production by new plants leading to an approximately static O_2 content. Conversely, increased CO_2 concentrations in the atmosphere, from fossil fuel consumption, could well act as an etherial fertilizer, promoting more plant life and hence more O_2 production.

Little is known about the stabilizing tendency of the oxygen cycle, but it appears that built-in instability does not exist. Later in the discussion of the effects of pollution, we will examine the broader impact of fossil fuel combustion on the earth's climate as well as on the oxygen cycle.

THE NITROGEN CYCLE

Even though N_2 is the most abundant gas in the atmosphere, it is unreactive and thus cannot be used directly by many organisms. Figure 1–7 describes the overall cycle, noting which nitrogen fixation processes are essential to

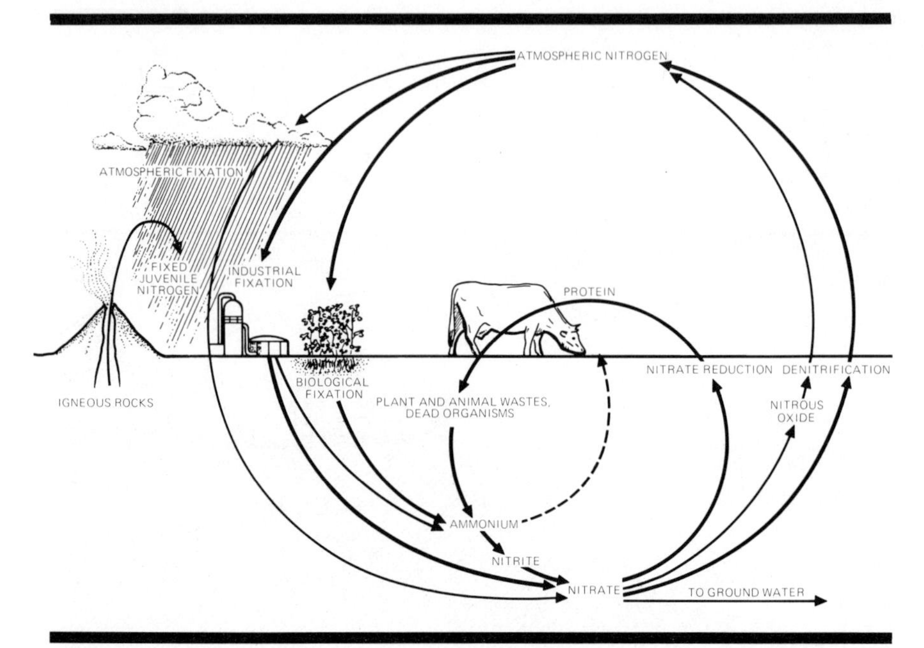

Figure 1-7. Nitrogen Cycle

Above is a simplified representation of the nitrogen cycle, in which nitrogen must be combined with hydrogen or oxygen to be assimilated by higher plants, in turn consumed by animals. Approximately 10 percent of increased nitrogen-fixing is accomplished by the large-scale cultivation of legumes and by industrial fixation, over amounts of nitrogen fixed by pre-agricultural terrestrial ecosystems.

microbial utilization. Fixation processes change inert N_2 into a usable form such as nitrate ion (NO_3^-), processed by plants into proteins and amino acids, which are in turn utilized by higher life forms (animals).

Nitrogen fixation is observable in the natural life cycles of bacteria (*Azotobacteraceae*) which convert N_2 into ammonia (NH_3) and other usable forms. Other bacteria (*Nitrosomonas*) convert ammonia into nitrite ion (NO_2^-) and so obtain energy for their own life processes. Still other bacteria (*Nitrobacter*) convert nitrite ion to nitrate ion (NO_3^-), a very effective source of nitrogen for plants.

This complex *life-based* nitrogen cycle is augmented by human nitrate production of an almost equal magnitude. Fixed nitrogen manufacture (made possible by the development of explosives around 1914) involves the reaction of N_2 with oxygen from air (to make nitrate) or with hydrogen from water or methane (CH_4) to make ammonia. Such materials are used extensively as fertilizers.

The other half of the nitrogen cycle is the denitrification process which is also carried out by bacteria. If it were not for these microorganisms, all N_2 in air might be in the form of NO_3^- in the earth's oceans. Denitrifying bacteria (*Pseudomonas denitrificans*) produce N_2 from oxidized forms of nitrogen, thus completing the cycle. Once again we see evidence of an indivisible biosphere providing its own "support system."

THE CARBON CYCLE

Life as we know it is based on carbon. No other basis exists on earth, although it might in some other planetary system. The carbon cycle depicted in Figure 1-8 shows the main flow of carbon-containing materials in the biosphere. The main cycle is the oxidation of carbon-containing materials into atmospheric CO_2, with photosynthesis returning it to the biosphere. The mechanics of photosynthesis and the possibilities of homeostasis were mentioned in connection with the oxygen cycle.

The most important features of the carbon cycle are:

1. It is almost perfectly balanced; the amount of carbon "fixed" annually by plants is oxidized back to CO_2.
2. There is a vast reservoir of carbon as carbonate (CO_3^{-2}) in sedimentary rocks such as limestone and dolomite. Indeed if the carbon in carbonates were somehow turned to CO_2 gas, the atmosphere would be hundreds or thousands of times denser than it is at present. Fortunately rocks change slowly, so we can be sure that this CO_2 will not be suddenly released.
3. Carbon is being modified by fossil fuel combustion, unquestionably resulting in increased atmospheric CO_2 concentrations.

OTHER CHEMICAL CYCLES

While water, oxygen, nitrogen, and carbon form the basis of life in the biosphere, there are myriad other elements that are also essential to life. One need only look at a box of breakfast cereal to see listed the "Minimum Daily Requirements" of iron, calcium, phosphorus, and other elements. Life forms require a wide variety of elements including sulfur, potassium, magnesium, sodium, and even such rare metals as manganese.

On geological time scales, even these elements are recycled in their own particular systems. Some of these cycles may be disturbed in important ways by human activity where the anthropogenic production greatly exceeds the natural amounts. There are two good examples of this kind of disruption.

Sulfur dioxide (SO_2) emissions from the burning of coal and oil are comparable in magnitude to the amounts emitted naturally from volcanoes; it is easy to recognize the resultant pollution problems.

Phosphorus is another mineral valued by man which is not abundant on the earth's surface. It is mined, used in detergents, and sent ultimately to the bottoms of lakes and oceans when waste water is discharged. Thus a precious mineral is literally thrown away in a manner that will make it difficult to retrieve when the mines are depleted in the next 100 to 200 years.

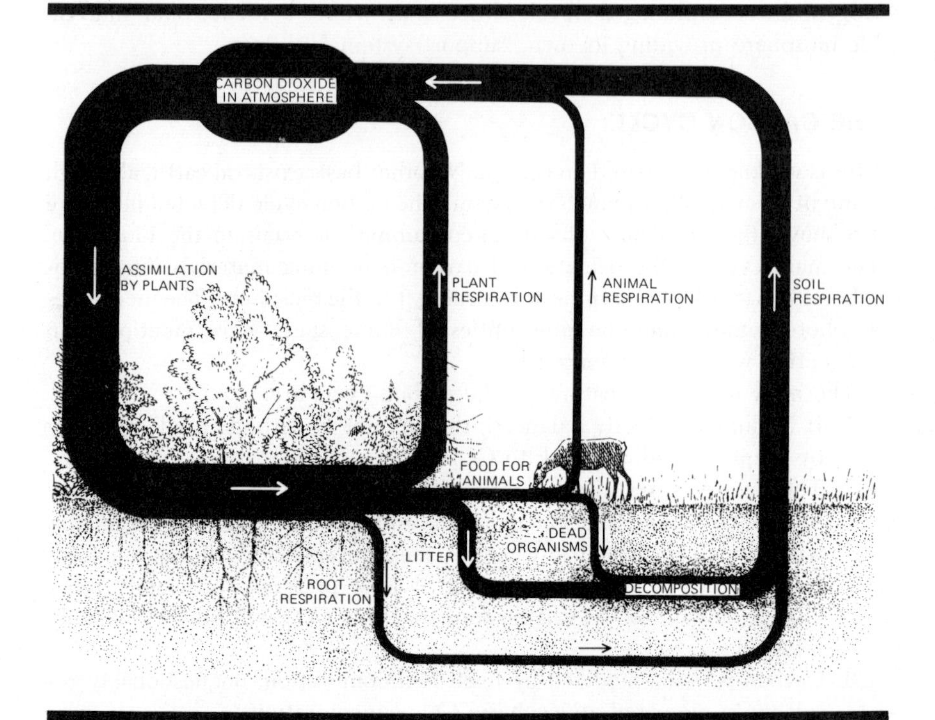

Figure 1–8. Carbon Cycle

Above is a simple diagram of the carbon cycle, which begins with photosynthesis. Carbon dioxide reacts with water to form hydrocarbons, releasing free oxygen to the atmosphere. Some of the carbohydrate supplies energy to the plant; carbon dioxide is released through the plant's leaves or roots. Some of the carbon fixed by plants is consumed by animals, which release carbon dioxide through respiration. Decomposition of dead plant and animal matter oxidizes tissue carbon to form carbon dioxide, which is returned to the atmosphere. The widths of the pathways indicate relative quantities involved at each stage.

Natural cycles are all intertwined; in fact, it is difficult to predict just what changes might occur in the one cycle from changes in another elemental cycle. The indivisibility of the environment thus seems to be one of the inescapable facts of our existence on earth.

SUMMARY

Man's current level of understanding of the natural world is capable of offering a hypothetical model of the formation of the earth. The model implies that the physical earth could have been formed as a result of purely random and natural physical and chemical reactions about 5 billion years ago. Similar reactions ultimately gave rise to certain complex, organic molecules which were capable of chemical replication and which produced free oxygen. As oxygen concentrations increased in the atmosphere, a wider variety of complex organic compounds was able to form randomly and interact outside a strictly aquatic environment. Thus the evolutions of the earth and of life itself are inextricably connected through a continuum of geophysical activity extending billions of years into the past. Current environmental conditions, such as the quantities and composition of our atmosphere and oceans, have probably remained remarkably constant for the last few billion years.

Although the overall compositions of our atmosphere and oceans tend to remain constant, each of the individual components of these systems is actively involved in dynamic equilibria, with continuous transfer of material and energy occurring between all systems. Gaseous components such as water vapor, oxygen and carbon dioxide, liquid water, and solid material such as dust particles, even the continents themselves are all involved in cyclical interchanges.

References—Chapter 1

1. Warren, Charles E.
Biology and Water Pollution Control.
Philadelphia: W. B. Saunders Company, 1971, p. 75.

2 Origin and Evolution of Life

CHEMICAL BEGINNINGS

Recognizing the sensitive relationships that exist between all life and its environmental surroundings is essential to an understanding of the significance of pollution. To develop this recognition fully, it is necessary to emphasize a few details regarding the probable mechanism of life's origin, namely its strictly chemical and physical beginnings, the dependence of the entire process on chance associations of many events, and the extremely long period of time that was required. It is difficult to perceive the immense time scales involved. Radioactive dating shows that the earth is about 4.5 billion years old, and the oldest known fossil evidence indicates that microscopic life was formed about 3 billion years ago.[1] Thus, the time between these ages can be regarded as the period of chemical evolution. See Figure 2–1.

We have already established that the early atmosphere was of a reducing nature; that is, molecular oxygen (O_2) was not present. We can surmise that oxygen did not appear in appreciable quantities until about 1 billion years ago, because only rocks younger than about 1 billion years contain oxidized iron.[1] About 0.8 billion years ago the atmosphere became fully oxidizing, meaning that evolution had proceeded for about 2 billion years in an oxygen-free (anaerobic) atmosphere.

In the earlier reducing atmosphere, gaseous molecules of methane (CH_4), ammonia (NH_3), hydrogen sulfide (H_2S), and water were present. The major elements that comprise biological matter are contained in this group of molecules: nitrogen (N), hydrogen (H), carbon (C), oxygen (O), and sulfur (S). At the same time, sources of energy existed that could have caused reactions of these compounds, producing more complex substances similar to those contained in biological material. These energy sources included ultraviolet radiation (UV), lightning, thermal energy from volcanic activity, and radioactivity. Natural conditions have been simulated in the laboratory in which the exposure of ammonia, methane, and water to an electrical current produced important biochemical building blocks such as amino acids, sugars, purines, pyrimidines and fatty acids. Phosphorus and sulfur can also be incorporated into compounds resulting from these reactions.

As the vast time periods unfolded, increasingly complex carbon compounds were formed through chemical reactions. In the presence of heat and light, chemicals produced randomly tended to group themselves into long chain compounds such as polypeptides and polynucleotides. For reasons which are not entirely clear, environmental factors operated to encourage the existence of some complex types and discourage others. One can imagine a long history of accumulation of different molecular assortments, particularly in inland seas, where wind and evaporation would concentrate them until, finally, some molecules existed that were capable of replicating themselves.

Scientists feel that local volcanic activity alone could have supplied both the substance and the energy (temperatures of lava can be 1000°C) for these condensation reactions. For example, in laboratory experiments powders of simple amino acids, the building blocks of protein, have been distributed onto volcanic rocks heated to a temperature of about 200°C. Upon contact, the powders accumulated into colored smears that were found to contain proteinaceous material. When water was added, many of these materials went into solution. The products of these artificially produced proteinaceous materials or proteinoids were not easily distinguishable from simple bacterial cells when viewed with an electron microscope. To be sure, scientists can only speculate about the mechanisms of these chemical reactions. It is commonly felt, however, that the complex chemical structures that we know to be essential for life today could easily have come from a long history of natural chemical reactions tempered by interaction with the environment.

From molecules of proteinaceous material to living organisms capable of reproduction is a big jump. As mentioned earlier, development of life would have required protection from the intense UV radiation, and therefore these early reactions probably occurred under water. The first organisms would have needed a mechanism by which they could gain energy from the environment and a hereditary mechanism to ensure their continued replication. To accomplish this, a cell structure would have been required to contain the

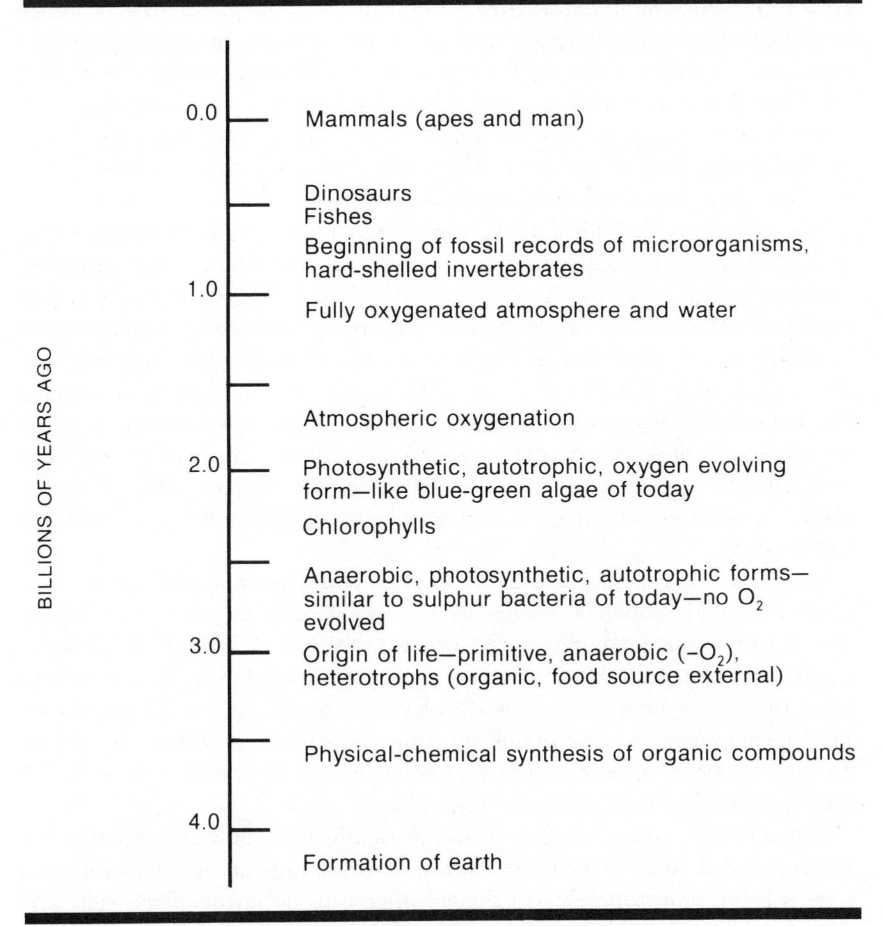

Figure 2-1. Important Events in the Origin of Life and Organic Evolution
After T. D. Brock, BIOLOGY OF MICROORGANISMS, Prentice-Hall, Inc., 1970.

complex chemicals that are necessary for performing the function of energy and nutrient gathering and cell replication.

CHRONOLOGY OF LIFE

What was the nature of the first living organisms and how would they compare with those that are present today? Since the early environment was oxygen-free and remained that way for more than 2 billion years, the first organisms were necessarily anaerobic. The thick soup of organic compounds likely formed an unlimited food supply for the first organisms, which were dependent upon such sources for energy and new carbon atoms for additional growth (heterotrophy). Later, after a cell wall and more complicated cellular machinery developed, organisms may have been present that would resemble the anaerobic sulfate-reducing bacteria of today. These are simple one-celled organisms that, without oxygen, could have oxidized the accumulated organic compounds to obtain energy for growth and other cellular functions. Thus, we have organisms living today that could have functioned easily in the rigorous environment of the primitive earth.

The enormous biological complexity and diversity that currently exists could probably never have developed without tapping a much more extensive source of energy than that provided by the chemical-physical reactions just described. The chance production of chlorophyll allowed microorganisms to utilize a nearly unlimited source of energy—visible light from the sun. The entrapment of light energy by chlorophyll in green plants to produce chemical energy needed for cellular functions is called photosynthesis. With the advent of this process, the production of organic material in the form of organisms increased greatly. Furthermore, this process allowed green plants to synthesize their own food supply from CO_2 molecules in the air (autotrophy).

Molecular oxygen (O_2) is a by-product of photosynthesis and this process is the principal source of oxygen in our atmosphere. Oxygen in the atmosphere and oceans gradually increased from the time the first photosynthetic organisms appeared, about 2 billion years ago, until 0.8 billion years ago when the atmosphere is considered to have reached a fully oxidizing status. This development is most significant, since an oxidizing atmosphere paved the way for development of aerobic organisms and increased organic evolution at a rapid rate, as shown in Figure 2–1.

The scientific record of the evolution of life is at best incomplete. We are dependent upon radioactive dating of fossil remains in rocks for such a record, but unfortunately soft-shelled organisms do not fossilize well. The record we do have begins about 600 million years ago with hard-shelled invertebrates such as brachiopod and gastropod molluscs and trilobites. Invertebrates dominated the scene for about 200 million years before the first vertebrates evolved. The emergence of life into the terrestrial environment began with the evolution of lung-breathing fish between 350 and 400 million years ago. Some species of lungfish exist even today. Frogs and toads are remnants of a greater variety of amphibians that lived 350 million years ago. Although able to live on land for long periods, they are not independent of the water environment, because their eggs have thin membranes which easily dry out unless bathed in water. Development of a hard-shelled egg permitted reptiles to enjoy total independence of the water. The often dramatized age of reptiles, dominated by the dinosaurs, lasted nearly 200 million years, from 280 to approximately 100 million years ago.

Reptiles were replaced by mammals, rather obscure and certainly small compared to the dinosaurs, but unique for several reasons. They were covered with hair and had a constant body temperature, and, most importantly, they provided care and food for their young. These features allowed mammals to adapt to a much more varied environment because, in a sense, they carried their environment with them. No longer was their metabolism at the mercy of changing temperatures; they could be as active in cold weather as in warm. Some scientists believe that it was the lack of a constant body temperature that led to the decline of reptiles. Another mammalian advancement was the early development of young inside the female. Over 100 million years ago the age of mammals was in full force, but the evolution of man was still a long way off.

What must be remembered is that these events occurred over millions of years, and that species as we know them changed only gradually. Hundreds or thousands of years passed before beneficial changes occurred in a species to afford it some differentiation from other species. We see some evidence in our own time of species change, but all in all it is a very slow process.

MECHANISM OF EVOLUTION

What mechanism allows evolution to occur, that is, to change one species to another, one large group of organisms to another; amphibians to reptiles to mammals and on to man himself? The theoretical basis for evolution, the most unifying theory in biology, was proposed over a century ago by Charles Darwin. He emphasized the process of natural selection: those species that were most fit or best adapted to their environment would survive environmental stresses. Those better adapted would reproduce more successfully and leave more progeny; they would be more successful in competing for food and space, and the more poorly adapted organisms would gradually diminish in numbers. What Darwin did not discover was the mechanism which caused members of the same species to have different characteristics.

The physical and mental make-up of organisms is controlled by structures within cells called genes. Genes are located on chromosomes, which are composed of deoxyribonucleic acid (DNA). Hair color, sex, the veination of an insect's wings, the urge for fish to migrate, the inclination to defend territory, and human intelligence are all determined by the arrangement of chemicals within the DNA molecule. These characteristics are passed on from generation to generation through processes of sexual and asexual reproduction and cell replication. Just as the chemical reactions that first produced life on our planet occurred randomly and resulted in the formation of large varieties of chemical structures, chemical reactions and replications of the genes change randomly from time to time. The DNA molecule that determines the characteristics of an organism replicates constantly within an animal's body, causing it to develop from a single cell into a complex of billions of cells, eventually forming the whole organism. Random replicating or copying errors are almost certain to occur, and these changes are called mutations.

It is ironic that the majority of mutations are detrimental; only one in thousands affords the organism some measure of increased success in adapting to the environment. Mutations occur naturally and at random, and would have as great a probability of occurring in one species at one time as they would in another species at another time, although they may also be caused by such events as the release of radiation from atomic blasts or discharge of toxic chemicals. Genetic variation is the basis of heredity and leads to evolution. Once the detrimental or beneficial changes have occurred, natural selection by the environment is superimposed upon organisms and, as Darwin pointed out, the fittest survive.

An excellent example of natural selection operating through environmental change occurred in England. The population of a certain species of moth was composed primarily of members whose body coloration was whitish-gray, although some darker-colored moths could be found occasionally. As industrial development accelerated and the air pollution from coal burning began to coat the forest vegetation with dark soot, a shift in the moth population occurred. The result was that, within 50 years, the predominant color of the moths shifted from light to dark, and today light-colored moths are as rare as dark-colored ones used to be. Why? Darker moths are more camouflaged on the blackened trunks of trees and more difficult for birds to see than their light-colored counterparts. Therefore, dark moths lived longer and produced greater numbers of offspring than the once favored white variety, which initially blended with the natural background in the unpolluted environment.

Another example of evolutionary change is the development of resistance to persistent insecticides by insects and fish.[2] Through continual exposure to organo-chlorine insecticides such as DDT, the resistance possessed by a small percentage of individuals in insect and fish populations became dominant. Most or all of the population eventually became resistant to the insecticide. With some insecticides this resistance developed within ten years. The genetic change that resulted in resistance in a few individuals was no doubt random, and may have occurred long before actual exposure, but once it existed in the presence of insecticides, selection operated in favor of the resistant individuals, and they contributed a progressively greater number of progeny to the population. Development of resistance in rapidly reproducing organisms, such as insects, would be expected to occur sooner than in more slowly reproducing organisms, such as fish. The more generations produced per unit of time, the faster resistant mutant genes will be selected and eventually dominate the population.

APPEARANCE OF MAN

What can be said about the origin of man? From the incomplete fossil record uncovered so far, it appears that man originated in East Africa. A climatic change on that part of the earth some 25 million years ago altered large land areas from jungle to spotty timber and broad grassy savannahs. In this environment, a common ancestor of man and the great apes evolved, one that was better adapted to walking upright, getting his food from the savannah, and using the sparse trees as shelter. The rugged existence on the savannah probably created a heavily competitive environment that strongly favored the evolution of prehumans that walked upright, hunted with weapons, and developed the cunning to outwit stronger competitors. Certainly their brawn was no match for the large predators of the savannah. Prehumans existed in East Africa as long ago as 2 to 4 million years and are called Australopithecines. Evidence indicates that these prehumans used tools and may well have possessed much of the reasoning power of modern man.[3]

A two-million-year-old ancestor that walked upright, used tools, and hunted with weapons may seem unbelievable, but estimates are being extended even farther back, as dating technology improves and new archeological finds occur. Actually, even a million years is only a blink of an eye relative either to the time that life has existed on earth (1/1500) or to the age of the macroorganism fossil record (1/300).

The story is not complete by any means, but over 2 million years ago there were hominid-type animals walking upright on the African savannah, using weapons and carrying out a predatory existence much the way later men were to live. Because the survival of these men depended upon their use of tools, reasoning power, and the strength they derived from socializing and banding together, selective pressure strongly favored any genetic changes that contributed to these characteristics. Thus, evidence clearly indicates that man has an evolutionary past not unlike that of lower animals. He evolved in and may well be best adapted to an 85°F, humid, grassy savannah with sparse trees and unpolluted air, water and land. Given sufficient freedom, people tend either to construct or migrate to such an environment. As a product of this environment, man is still subject to the same environmental stresses, although they do not currently affect man's evolution in the same way that his predecessors were affected. Because numbers of human progeny are currently produced for cultural and social reasons, rather than to ensure survival of the species, selection in modern man is not effective in improving his genetic stock and adaptiveness, although he could conceivably degrade it for environmental stresses can cause genetic impairment.

The problems besetting man regarding the degradation of his environment are perhaps more acute than they have ever been, yet for the first time we are in a position to understand and control our future evolutionary development. The numbers of surviving offspring are no longer controlled solely by natural selection, but rather by cultural and social customs which unfortunately contain little regard for environmental compatibility. One can only hope that the human community becomes and remains sufficiently aware of its basic biological dependence and inheritance so that our customs do not lead us to catastrophe.

SUMMARY

The development of complex assortments of organic molecules able to function in the physical environment as "living entities" first occurred on the earth over 3 billion years ago. Many species were formed in the seas as a result of random chemical and physical reactions which still occur today, and, for reasons which are unclear, certain types of organic molecules survived environmental stresses and emerged through vast periods of geological time as the basic building blocks of all living material.

From these random and relatively simple beginnings, life forms have greatly increased in complexity and diversity. During at least the last half of the earth's existence, the development of life has had a major role in shaping the characteristics of the physical environment, including not only terrestrial and aquatic environments, but also the chemical composition of the atmosphere. Mutations or random copying errors during cell replication provide the material for evolution; only a very few afford a species any measure of success in increasing its survival under environmental stresses. The mechanism of evolution works to improve the adaptability of a population to a changed environment through the process of natural selection.

The appearance of man on the evolutionary scene is a relatively recent development. The incomplete fossil and archeological record indicates that the warm and humid East African savannah was the cradle of the human species several million years ago. Unlike any earlier form of life, the human species is capable of understanding the manner of its own development, and is therefore potentially able to control all future evolutionary development on the planet.

References—Chapter 2

1. Brock, T. D.
Biology of Microorganisms. Englewood Cliffs: Prentice-Hall, Inc., 1970, p. 495.

2. Ferguson, D. E.
"Mississippi Delta Wildlife Developing Resistance to Pesticides," *Agricultural Chemicals*, September 1963.

3. Boughey, A. S.
Man and the Environment: An Introduction to Human Ecology and Evaluation. New York: The Macmillan Company, 1971, p. 472.

3 Concepts of Ecology

DEFINITION OF STUDY AREA

If man's existence is one chapter in the total story of life's response to environmental pressure, then it follows that man is inextricably connected to his present physical and biological environment. Unlike any organism preceding him, however, man is effecting control over the environment and thus his own evolution. To illustrate man's intricate connection with his environment, it is useful to describe the most important concepts of ecology as they pertain to all life.

Ecology is defined as the interaction of organisms with their chemical-physical environment, or the interaction of the biotic and abiotic environments. An ecosystem may be either self-contained, such as a stoppered laboratory flask filled with pond water containing microscopic plants and animals living in equilibrium, or open, such as a small lake with water inlet and outlet. A more precise definition of ecology indicates that it is the study of the structure and function of such ecosystems.[1] The ultimate ecosystem is the earth or biosphere, which is self-contained like a rocket, hence, "the space ship earth."[2]

The term "structure" of an ecosystem refers to the way biotic and physical environments are organized. Within each ecosystem there are necessary components, usually plants, that form the base of the system and provide food energy, animals that use this energy to live and grow, and bacteria and fungi that decompose dead organisms and return the valuable chemical materials for reuse in the system. These chemical materials are principally carbon, nitrogen, phosphorus, oxygen, hydrogen, and sulfur—the same important elements involved in the origin of life. The energy source for the typical ecosystem is the sun. Solar radiation is utilized by green plants to form the initial energy for the system. In addition, there are physical stresses to which life in the ecosystem must adapt, such as the intensity and duration of sunlight, extremes of temperature and humidity, water currents, and soil composition.

STRUCTURE AND FUNCTION IN ECOSYSTEMS

The structure and function of ecosystems are interconnected and therefore difficult to separate in discussion. Their structures are often determined by the food-gathering mechanisms: in any ecosystem, there are those members which convert incoming sources of energy into an organic food supply, used by other organisms unable to tap the primary energy source directly. At each successive step in the energy transfer process from simple to more complex organisms, there is a loss of energy. When an animal eats a plant, far more energy is lost than is retained, and as a result, fewer numbers of more complex organisms exist at each level. Therefore, it is not surprising that organisms functioning as the primary converters of basic energy occur in the greatest numbers, and those organisms that are last to receive food energy in the system occur in the least quantity.

The transfer of energy through an ecosystem is referred to as a "food chain": food energy trapped by producer organisms is transferred to consumer organisms. Herbivores devour the producer plants and are in turn consumed by carnivores. Two simple food chains are illustrated in Figure 3–1, but seldom does this process occur as a simple three- or four-link chain. Rather, the organisms located at various levels exhibit a variety of feeding habits and thus are not solely dependent on members immediately below them, as would be the case in a simplified unidirectional food chain. More accurately, the structure for food energy transfer in an ecosystem takes the form of a "food web," with organisms at each step playing dual roles and interacting with each other through several interconnected links. The interconnected pathways can vary seasonally and from place to place as organisms age and change their diets, and as the type and quantity of food change. Man, for instance, is a part-time carnivore and part-time herbivore, although he is somewhat unique in that he occupies the top of the food web and has no effective predators. This point is of great significance and will be referred to again.

To illustrate the idea of a food web, let us consider more precisely how food energy might be transferred in a pond ecosystem. As shown in Figure 3–2, there are several organisms involved in transferring food energy from producer algae to carnivorous consumers. Even here many steps are left out, as plants, in addition to algae, contribute to the base energy. Energy in dead organisms can be reused to some extent by consumer organisms which feed on the bacteria and fungi that decompose the dead material.

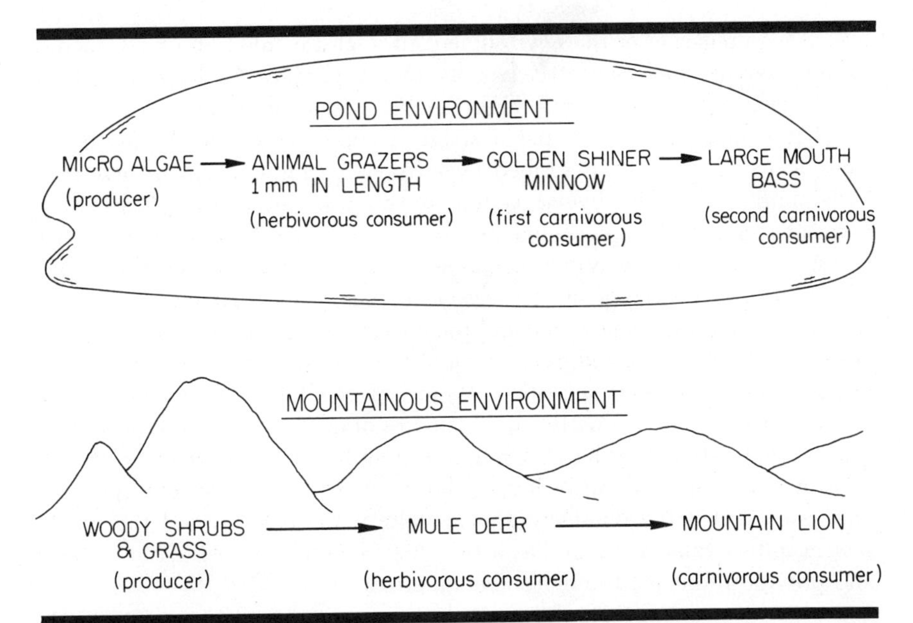

Figure 3–1. Simple Food Chains

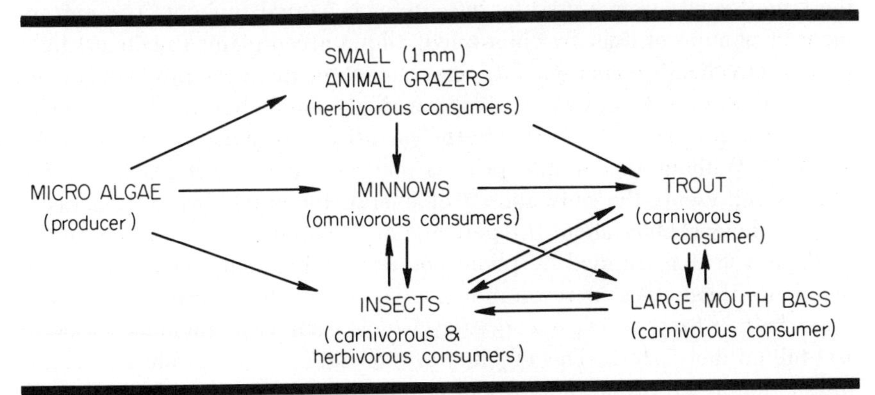

Figure 3–2. Simple Food Web

STABILITY IN ECOSYSTEMS

We are interested in the structure and functioning of ecosystems because, when maintained in a highly evolved and developed state, they represent finely tuned and stable systems which provide the most efficient use of solar energy and furnish man with his most dependable food source. The use of biologically fixed energy from natural systems must be kept in harmony with the surplus yield, or overuse will occur. Overuse may deplete one group of organisms and may lead to reduced efficiency and stability in the overall system. Simplified unstable ecosystems can also result from man's uncontrolled manipulation of the environment through the addition of extraneous toxic materials. Such activities tend to revert complex food webs back to simple food chains, changing a highly stable system to an unstable one.

What is meant by stable and unstable, complex and simple ecosystems? Complexity can best be described by introducing the concept of diversity. Very simply, a highly diverse system is composed of many species, each more or less equally represented. In contrast, a system lacking diversity has few species with as few as one species accounting for most of the individuals. Thus, one can picture a diverse system as being more balanced, with no species playing a dominant role and the energy in the system being supplied equally to a variety of different species. A great deal of "information" can be said to exist in such a system. If one species falters, many others are available to assume its function and little loss of efficiency in the total system is suffered. On the other hand, a simple system lacking diversity is relatively unbalanced. It operates according to the functions of a few species (perhaps only one), and if the dominant species falters, the efficiency of the whole system suffers. Such a system has little "information" from which to operate and provide resilience to environmental stress.

As we have already emphasized, the structure of ecosystems is closely connected to the source as well as the amount of energy available. By far the largest single source used by biosystems is natural sunlight. The entrapment of photons of light by chlorophyll allows green plants to convert light energy into chemical energy. With light energy the plant can produce glucose from carbon dioxide and water. This seemingly simple but highly significant reaction is the necessary basis for the quantity and diversity of life as we know it. Without this simple process man could not exist. Although the process represents the only source of energy for organisms, it is actually very inefficient. Only about 0.1 percent[3] of the total sunlight falling on the earth in a year is trapped by photosynthesis, and of that, half is consumed by plants themselves in respiration and is given off as heat energy. The other half is fixed as organic matter, which each year amounts to about 164 billion metric tons. This quantity is referred to as net production (gross production minus respiration). About two thirds is contributed by land plants and one third by microscopic plankton algae in the ocean. A small amount of the net production may be stored in sediments and forests, but the biospheric cycle of organic matter is essentially in equilibrium. Nearly all organic matter that is produced is converted back to the original constituents—CO_2 and H_2O—and the energy is converted to heat by consumers and decomposers.[3]

How much of the net production of organic matter from sunlight can man acquire, and what laws guide his quest? As the energy trapped by chemical bonds in food is transferred from producer plants through the food web, extremely large losses occur. These losses amount on the average to about 90 percent, resulting in a clean conversion of only 10 percent, as shown in Figure 3–3. Thus, the more steps in the transfer process, the greater the reduction in the energy available for use. For example, man would obtain ten times more energy by eating a fish that fed exclusively on producer algae than by eating a fish that ate a smaller fish. In the first instance, 10 percent of the original energy would be available, while in the second the figure is only 1 percent. Man also faces the same decision when he chooses to eat beef rather than the grain required to produce the beef.

To realize why these energy losses occur requires some understanding of the laws of thermodynamics. The first law of thermodynamics states that energy everywhere is conserved; it can neither be created nor destroyed.

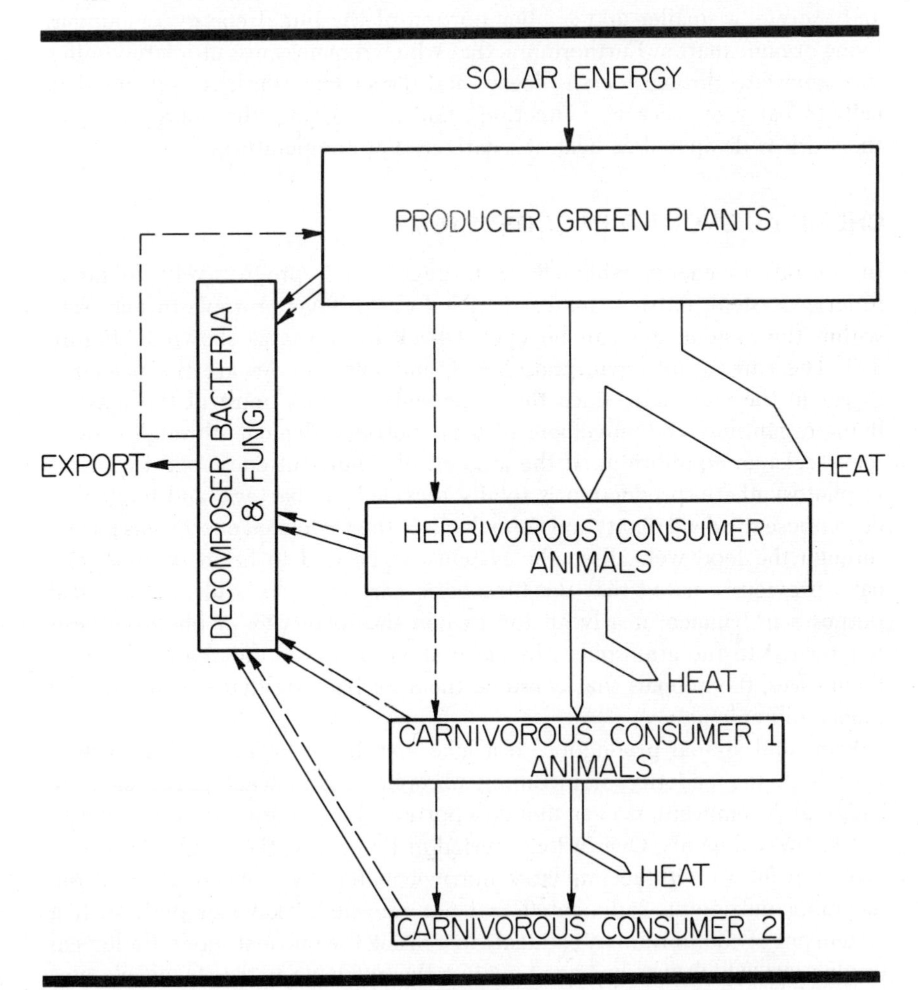

Figure 3–3. Energy Transfer and Nutrient Cycling in an Idealized Ecosystem

From this law we know that the total energy initially fixed in the system must equal the total energy remaining after transfer through the system. However, little of this energy remains in the form of chemical bond energy in living matter; the bulk of it, 90 percent at each step, is converted to heat energy and lost to the environment. The second law provides the real explanation for inefficiency in the energy transfer system. This law refers to the direction in which the process proceeds, and states in essence that energy will inevitably disperse from high concentration to low concentration. Regardless of the form, it will seek a state of randomness in nature, from the most concentrated and usable state to the most dispersed, unorganized and useless. Thus, as organisms at each step in the food web consume, respire, and excrete, a smaller and smaller portion of the initial energy remains as living organic matter. Furthermore, that which remains must ultimately suffer the same fate through death, decay, and dispersion. Work is performed in cells to carry on necessary functions, and in so doing, the energy used in the work is dissipated as heat at relatively low temperatures.[4]

CHEMICAL CYCLING IN ECOSYSTEMS

In contrast to energy which flows through and is progressively degraded in an ecosystem, nutrient materials that form cellular protoplasm may stay within the system and can be cycled back for reuse as shown in Figure 3–3. The carbon, nitrogen, phosphorus, and sulfur cycles are the principal cycles in the biosphere, since these elements compose most of the mass in living organisms. The movement of these nutrient elements through ecosystems seeks an equilibrium. If the amount of a nutrient consumed by plants in photosynthetic production is totally recycled (by bacteria and fungi that decompose the dead plants and animals, after the associated energy has passed through the food web), then the system can be said to be in balance. We have previously noted that the biospheric carbon cycle is for all practical purposes in balance; nearly all the carbon that plants fix in photosynthesis is returned to the atmosphere by respiratory processes either in the plants themselves, the animals that consume them, or bacteria that decompose the plants and animals.

Plant and animal production in a lake may be in long-term equilibrium with both the quantity of incoming phosphorus and what is recycled by bacterial decomposition, even though a portion of the phosphorus is constantly lost to the sediments. Over a long period of time—say, thousands of years—lakes can fill up with accumulated nutrient material if some fraction of the incoming nutrient is sedimented and not recycled. Man can push such a system out of equilibrium by suddenly increasing the nutrient input. Biological production, which was balanced against the available nutrients, accelerates because of the increased food supply. Sufficient time is not available for the biosystem to reach equilibrium with these nutrients, and nuisance conditions from excess, unused production result. In essence, this superabundance of nutrients is the cause for the problem of eutrophication, or excessive production and accelerated aging in water bodies. By reducing man's contribution of nutrients and allowing the input to return to a natural level, the system could be expected to return to equilibrium in a reasonably short time.

Biogeochemical cycling is involved in the ecological effects of virtually all substances that man adds to ecosystems. Many deleterious materials other than nutrients have an affinity for biological accumulation and can be "magnified" in the food web. That is, organisms take up the materials and concentrate them at higher levels than exist in the surrounding environment. The material either simulates the action of a biologically required material or is easily absorbed through cell walls and membranes of plants and animals. Examples of these processes are the replacement of calcium by strontium-90 in bone tissue and the cellular concentration of fat-soluble organochlorine insecticides. This biogeochemical process is favorable to man in that nutrients are concentrated in the food web, allowing easy harvesting, considering the work that would be required if he were to set about accumulating those nutrients in a synthesized form without the aid of a food web. The same process can also magnify toxic chemicals in the environment, and can result in man's obtaining an excessive, possibly lethal, dose of the material from animals.

As was previously stressed, life evolved as a result of a physical and chemical environment whose quality has remained relatively constant for billions of years. One of the finest examples of the effect of life itself on the shaping of the environment is the occurrence of molecular oxygen in our water and atmosphere as a result of the photosynthetic activity of green plants. In a similar way the cycling of nutrients, such as carbon, nitrogen, phosphorus, and sulfur, is interrelated with and dependent upon the structure of the ecosystem food web. Man can seriously upset the rhythm of ecosystem nutrient cycling on a large scale, either by inflicting direct mortality on less tolerant species or by slightly changing the environment so that these species are lost in the natural selection process. Both accomplish the same end; only the time required is different. As mentioned earlier, the CO_2 content of our atmosphere is increasing at about 0.2 percent per year.[5] Because CO_2 is the principal source of carbon used by plants in photosynthetic production, such changes may eventually alter organic structure and function. Many species of terrestrial plants tend to be limited by the scarcity of CO_2. Increasing the quantity of CO_2 would stimulate their growth.

We can really only speculate on the extent to which drastic alterations of nutrient distribution would affect cycling patterns on a large scale. However, man is manufacturing fertilizer, force-growing mono-species crops, transporting these products with their incorporated nutrients to urban centers, and releasing the wastes into the ocean through sewage treatment plants, thereby introducing totally artificial and unlinked cycles. Whether these massive operations are having an effect is largely unknown, except in intensively studied and isolated areas. A significant and possibly irreversible effect may take a long time to become detectable. Certainly one can imagine how such an artificial cycle without proper management could ultimately result in increased plant and animal growth in areas where man is unable to harvest it and otherwise keep the nutrients in a useful cycle.

EFFECT OF STRESS ON ECOSYSTEMS

Now that we have introduced the concepts of structure and function of ecosystems, let us combine these concepts and characterize natural, degraded, and controlled ecosystems. As we stated earlier when discussing structure, man is or should be very much interested in how and why the properties of ecosystems change, and in the consequences of change. Remember that an ecosystem tends to evolve toward complexity and diversity, involving many species that are each of relatively equal quantitative importance. To simplify an ecosystem, or to force it back to a younger stage, requires work, since its natural tendency is to gain complexity. By gaining complexity an ecosystem also gains a fairly large quantity of organisms or biomass and becomes stable. This stability can be attributed to the fact that much of its energy is tied up in biomass; productivity per unit biomass is low and the yield of organic material on a sustained basis is also low.[1] This means that fluctuations are minimal, since productivity per unit biomass or the percentage change in the observable biomass per unit time is small. (See Figure 3–4.)

A degraded ecosystem is somewhat the opposite of the natural state. Diversity is low; that is, there are few species, but their numbers are great. Production per unit biomass is great, and since there are few species the yield or material available for capture in a single species on a sustained basis is large. Of course, with the high ratio of productivity to biomass, the percent change in observable quantity of organisms or biomass with time is great. Therefore, such an ecosystem is considered unstable because fluctuations tend to be large, as shown in Figure 3–4. These consequences can result from insults, whether natural or man-caused, inflicted upon natural ecosystems. Altering nutrition by adding sewage, causing subtle or dramatic mortality by adding toxic agricultural or industrial wastes, or damage from lightning-caused fire all tend to select for more tolerant species and simplify the ecosystem.

A controlled ecosystem is similar to a natural ecosystem in that it tends to be relatively stable. However, it resembles a degraded ecosystem in that its structure is simple. A cultivated field is a controlled ecosystem. The biomass is composed of only a few species, usually one, and it is productive

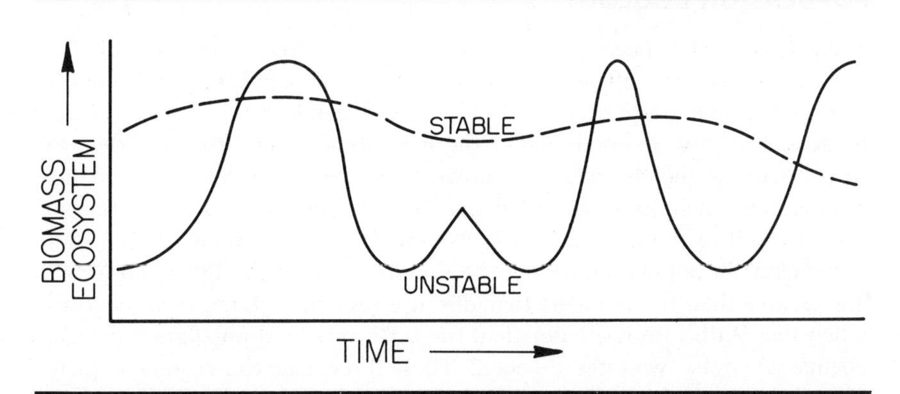

Figure 3–4. Biomass Fluctuations in Stable and Unstable Ecosystems

because the energy is funneled into the one species which is desired by man. The sustained yield is great since most of the energy is directed into a single species crop that is available and usable. On the contrary, if much of the energy is directed into weeds and insects that are not available or usable, the yield is considerably reduced. Such a relatively stable and productive ecosystem is not in equilibrium and requires exceptional control (work) to maintain it in that state. Since this maintained stability is unnatural, there are risks involved with highly controlled (agricultural) ecosystems. Water and fertilizers are usually required to maintain stability and productivity. Because the crops are monospecies and usually quite succulent, they are highly susceptible to insect pests and diseases. The entire biomass could be eliminated by a single insect outbreak, since there are no reserve species. Protection against such an event must be maintained by frequent treatment with pesticides. Thus, the risks involved in ecosystem control are twofold. In the first place, constant vigilance is required; any laxity in control could result in deterioration of the entire system. Secondly, the side effects on natural ecosystems from the use of insecticides, fertilizers, and water can be extremely costly. These effects will be discussed in more detail later. The increasing risks are real because fertilizer use is expected to double and pesticide use to increase six times by 1985 in order to feed the world's burgeoning population.

As Odum[1] has stressed, man should seek a compromise between highly controlled ecosystems of "production simplicity" and totally natural ecosystems of "protection diversity." To attain this goal of moderate diversity and productivity in reasonably stable ecosystems requires much greater understanding of the structure and function of ecosystems. However, such a goal is desirable if man is to maintain ecosystems that will serve his many needs and still be reasonably stable and resistant to shock.

POPULATION ECOLOGY

The existence of a species in an ecosystem is obviously not as one individual but as a population of individuals. When several populations live close to one another and are partially or totally interdependent, a community is said to exist. Stability or instability in the ecosystem results from the diversity of populations and the relative increases and decreases in the size of those populations. Changes in the total quantity of biomass in an ecosystem are then a result of the net changes in its composite populations.

Increases in population occur when numbers added by birth and growth are greater than the numbers that die in a unit time. Population decreases when the death rate is greater than the birth rate, and population does not change when the two rates are equal. These three conditions can be depicted for a hypothetical population which has a constant birth and death rate, as shown in Figure 3–5. The difference between birth and death rates is termed the growth rate and is considered constant for the hypothetical populations whose trends in biomass change were illustrated in Figure 3–4. In nature, of course, the rates would not necessarily remain constant. A constant rate of increase in an animal population is similar to the pattern of increase in a savings account that grows at a constant percentage rate of interest

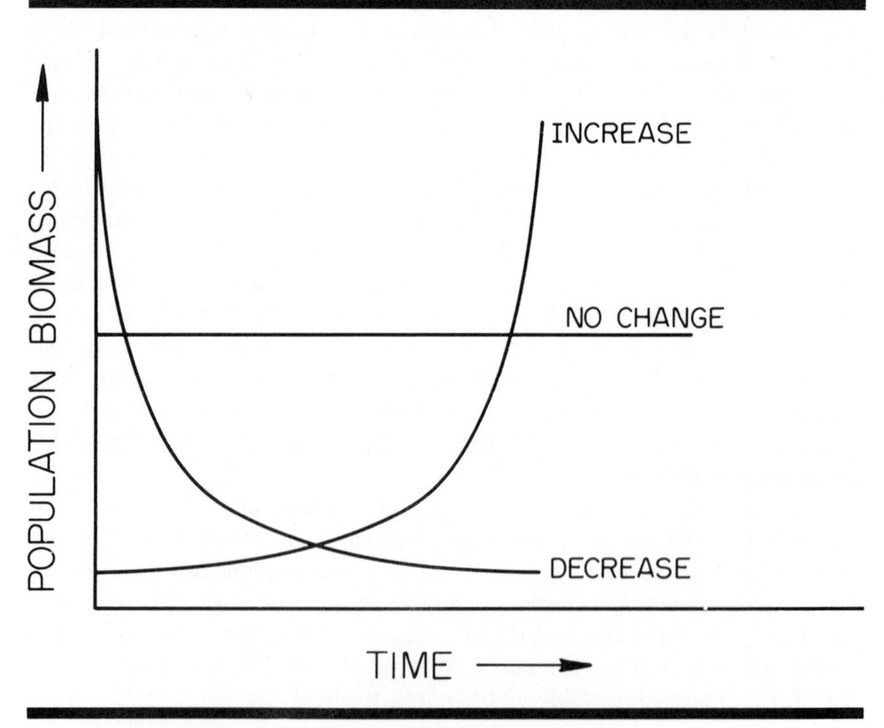

Figure 3–5. Possible Changes in Population Density With Time

on the new balance each time it is compounded. This type of increase or growth is called exponential—the population or savings account does not simply grow by some percentage of the initial balance or population but by a percentage of the existing total. Thus, one can see that as the account or population grows, the rate of growth increases, and, instead of the population increasing by some constant increment, with time the increment itself increases.

The growth and death curves depicted in Figure 3–5 are characteristic of uncontrolled rates of change in any population but can be most easily observed with microorganisms that have simple life cycles in which all members reproduce. Microorganisms can double their populations each generation by simply dividing once. With sexually-reproducing organisms, such as man, there is an age and sex structure in which only a portion of the population contributes to the growth of that population. As organisms increase in size and their life cycles become more complex, their biotic potential for increase, the maximum possible growth rate under ideal conditions, tends to decrease. Thus, a bacterial population that divides every 20 minutes could easily increase to over 1 billion cells in 10 hours starting from a single cell. A protozoan is slightly larger than a bacterium, requires 1 day to divide, and could produce a population equal to the volume of the earth in 3 months time. Such population growth is uncontrolled—it increases according to the maximum inherent potential. In reality, populations obviously do not sustain their maximum possible growth for very long. Environmental pressures retard their growth quickly, the maximum growth rate is reduced or exceeded by the death rate, and the population declines or stabilizes. A principal check on uncontrolled growth is predation. When the cougar population on the Kaibab plateau in Arizona was depleted in the 1930s, in the name of predator control and livestock protection, the deer population responded with an increased rate of growth. The actual birth rate of the deer probably did not change greatly, but with the removal of cougars, the principal predator, the death rate was greatly reduced. The deer population could not continue indefinitely at such an expanded growth rate—some environmental requirement was destined to run out. When food supplies became exhausted, the deer population collapsed.[6]

Internal controls provide another check on population. As the population density of some animals such as lemmings and hares increases, anxiety and stress cause reproduction rates to decrease. Parental neglect results in lower offspring survival for the African wildebeest when population density increases, and for the African lion when its food supply is low.[7] To survive, young wildebeests must find their mothers immediately after birth, a task which becomes increasingly difficult as the size of the herd increases and confusion spreads. Young lions are the last to feed, and during droughts and when food is scarce, the young are the first to starve. These and other methods, even including infanticide among primitive tribes, maintain population stability. It seems clear that greater stability results as food webs become more complex providing more interconnected prey-predator relationships.

Is man also controlled by internal as well as external forces so that there is a limit to his population growth and size? We have discussed man's probable origins and evolution, and it seems clear that he is dependent, as are animals, on food webs for his survival. Thus, there is no reason to expect that human populations should not be subject to environmental factors and stresses similar to those of other organisms.

SUMMARY

Particular groupings of life forms and the environmental segments they interact with are known as ecosystems. Ecology is the study of the structure and function of ecosystems, or, to put it more generally, ecology is the study of the interaction of living organisms with their chemical and physical environment. The structure of ecosystems is largely determined by patterns of energy-gathering mechanisms that are beneficial to all members. In general, solar energy is concentrated by photosynthetic organisms into chemical compounds which can be used for new growth. Energy stored in this solid form is available to higher organisms, although large energy losses occur with each consecutive transfer. Energy, initially trapped by green plants, is eventually returned to the environment in diffuse and unusable forms through heat losses and decay of dead organic matter. Without continual renewal of energy, a system would run down. As a result of energy transfer inefficiencies, primary producer organisms are present in the greatest numbers in any ecosystem. Nutrient chemicals that comprise the biomass and store energy can be continually recycled. The overall structural pattern by which energy is transferred in a complex ecosystem is referred to as a food web.

The term "food web" implies a diverse energy transfer pattern with many species providing the functional redundancy that is characteristic of stable and advanced ecosystems. Such systems are healthier in the sense that they are more resistant to gross perturbation and are more resilient under stress. If left alone, natural ecosystems tend to become complex, diverse, and thus, more stable. Under these conditions the existing biomass is rather large, but the production of new material per unit of existing biomass is small. Degraded ecosystems, on the contrary, have a low diversity and the observable increase in new material per unit of existing biomass may be large. Such an ecosystem is considered unstable because fluctuations in biomass tend to be large. Degradation of a stable ecosystem may result from natural or man-made insults such as the alteration of nutrients by the addition of sewage, the addition of toxic agricultural or industrial wastes, or the incidence of lightning-induced fire. All such insults tend to select for more tolerant organisms within the ecosystem and thereby simplify it.

Biological production in an ecosystem is a function of the growth and death rates of populations of organisms, and therefore occurs when the growth rate is greater than the death rate. The net effect of insults on the structure and function of ecosystems is the sum of the responses of individual populations to those insults, which may range from all populations growing near their maximums to the complete elimination of all populations.

References—Chapter 3

1. *Odum, E. P.*
"The Strategy of Ecosystem Development," Science, Vol. 164, 1969, pp. 262–270.

2. Boulding, K. E.
"The Economics of the Coming Spaceship Earth," in *The Environmental Handbook.* New York: Ballantine Books, 1970, pp. 96–101.

3. Woodwell, G. M.
"The Energy Cycles of the Biosphere," *Scientific American*, Vol. 223, 1970, pp. 64–74.

4. Ehrlich, P. R., and Ehrlich, A. H.
Population, Resources and Environment. New York: W. H. Freeman and Co., 1970, pp. 54–55.

5. Roberts, W. O.
"Man on a Changing Earth," *American Scientist*, Vol. 59, 1971, pp. 16–19.

6. Kormondy, E. J.
Concepts of Ecology. Englewood Cliffs: Prentice-Hall, Inc., 1969.

7. Ardrey, R.
The Social Contract. New York: Atheneum, 1970.

4
Waste Production of Civilization

INTERACTIONS OF WASTES IN THE ENVIRONMENT

We will now examine the various activities which man has developed to provide for his existence and for the quality of life he desires. Next we will examine the wastes and by-products which result from these activities. The goal of this analysis is to specify the types of waste that are produced and that will eventually have to be returned to the environment. Man's major activities include supplying food, water, and air for himself; developing shelter and living space; providing mobility; and searching for a certain aesthetic quality of life. Each of these activities produces waste which upon disposal interacts with the principal environmental elements: air, water, and land.

Throughout living systems, energy is prerequisite to function, and life seeks to preserve itself. Organisms continually strive to acquire the materials necessary to construct new cells, and must couple cell synthesis and energy production in order to survive. The production of new cells and energy involves restructuring molecules and releasing unused portions, molecules no longer needed, or oxidized compounds stripped of some of their energy. Thus, maintaining life requires the production of wastes. The metastable waste compounds normally can be incorporated into the life cycle of some other organism so that compatible living groups can maintain continuous flows of materials.

With his continual quest for insulation from his natural environment, man has produced ever-increasing amounts of products alien to environmental cycles. Once these products are released, the problem of making them assimilable in the natural flow of materials becomes significant.

Man, in small numbers, can be tolerated as a parasite in the biosphere. When man's numbers and activities occupy a significant portion of the biosphere, the problems of waste assimilation, and even continued life, become paramount. Building a home requires taking over land space, destroying trees for lumber, wasting soil, destroying habitats, and so forth. Every new *production* involves *destruction*; thus, man must evaluate the alternatives: whether the product is really more valuable than what is destroyed.

All the processes we have discussed involve consideration of the mass balance shown in Figure 4–1. As shown in this figure, air, water, energy, and raw materials provide products, by-products, and wastes. Recycling and process renovation provide opportunities for conserving materials within the system. By-product recovery or improvements in product recovery can substantially reduce the quantity of material entering the waste stream, but all processes eventually produce some waste as an inevitable part of production. Note the expanded mass balance in Figure 4–2. Note also that in terms of the mass balance, at steady state, the total pounds of material entering the system must equal the total pounds of material leaving the system, and the total energy entering must equal the total energy leaving the system. The mass balance shows that the production system is not an unfillable sump for materials and energy but rather a flow-through process, in which what enters the system must either exit from or accumulate in the system.

The satisfaction of particular needs in man's life takes the form of various industrial, agricultural, or other professional organizations as generally outlined in Table 4–1. Each of these activities is subsequently supported by sub-industries such as the chemical, petroleum, and fertilizer industries, and

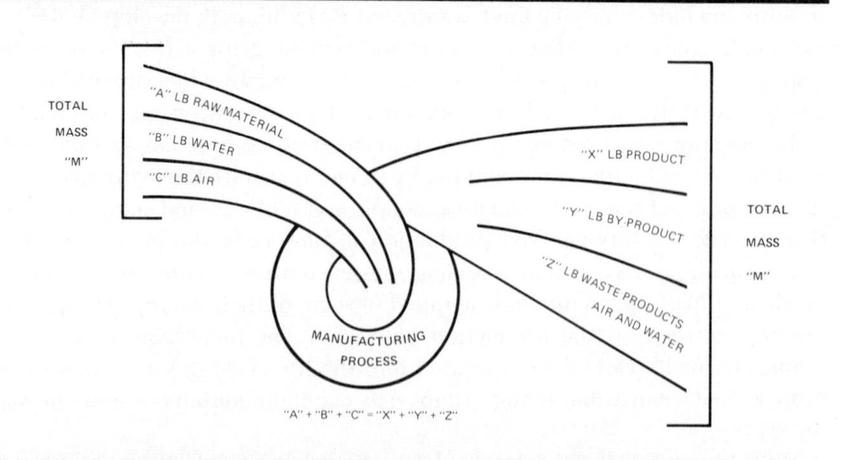

Figure 4–1. Mass Balance

each of these in turn creates waste. Disposal of these wastes can be made to any one of the elements of the environment.

The natural interactions between wastes and the physical environment are indicated by Figure 4–3. In this illustration, there are 11 processes shown which will interact and circulate wastes in the earth's environment. Table 4–2 summarizes each of these natural processes which act as mechanisms to transfer a waste between environmental elements. These processes are extremely complex, and there are a multitude of effects which can arise

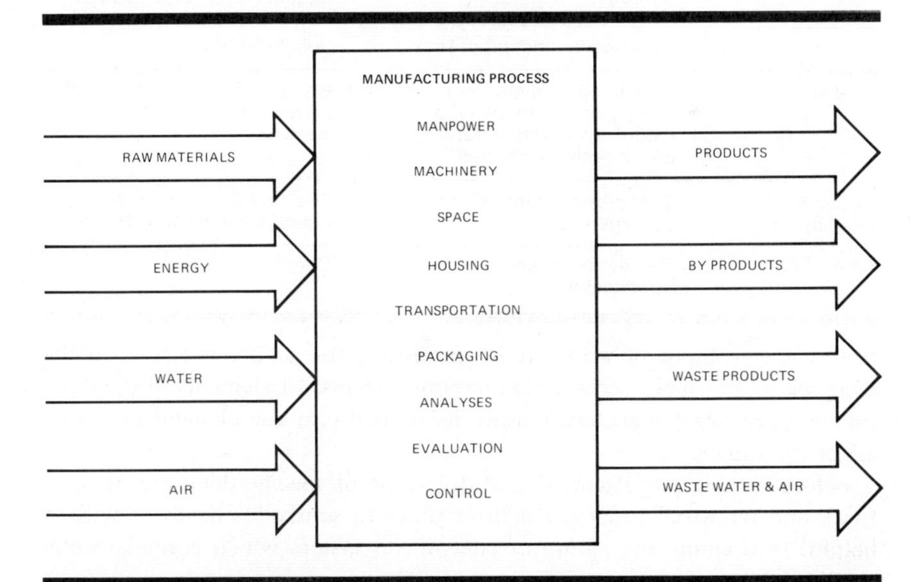

Figure 4–2. A Typical Manufacturing Process

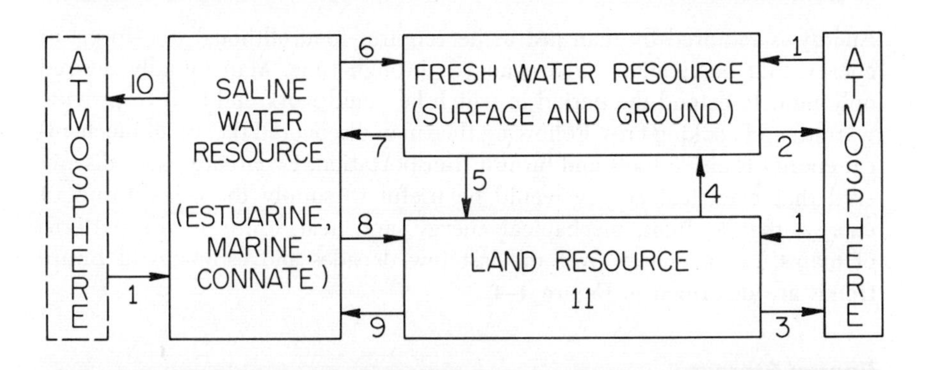

Figure 4–3. Interrelation of Environmental Segments

From WASTE MANAGEMENT AND CONTROL. Publication No. 1400, Committee on Pollution, National Academy of Sciences, National Research Council, Washington, D.C., 1966.

TABLE 4–1

Wastes Created by Satisfaction of Man's Needs

NEED	ACTIVITY	TYPICAL WASTES
Food	Agricultural and Food Processing Industry	Food covering, excess water, spoiled foods, packaging wastes, treatment chemicals
Water	Water supply, regulation, control	Excess water, solids, waste treatment, chemical sludge
Air	Air conditioning processes	Solids, heat, water
Shelter	Construction, materials, power, flood control, land modification, Regulatory and Housing Agencies	Waste materials, transportation wastes, land and landscaping wastes
Space & Mobility	Transportation industries and agencies	Fuel and transportation construction wastes, solids
Aesthetics	Recreation, scenic appreciation	Debris

from a single input of waste. It is interesting also to observe that, in the complex interchanges occurring in natural systems, all elements of the environment are interconnected. Wastes discharged into one element may well affect the others.

Before turning our attention to a discussion of specific domestic, agricultural, and industrial processes man employs to satisfy his needs, it will be helpful to examine the nature of energy conversion, which is fundamental to all material processes.

ENERGY CONVERSION WASTES

Energy is required by man just as he requires food, although the forms of energy that he requires have changed through time. Man initially desired only light to extend the period in which he could work and heat to provide warmth and cooking fires. Following this, man discovered the use of mechanical energy to drive tools and furnish transportation. Eventually, man discovered that electrical energy would be useful to supply the other forms of energy, that is, light, mechanical energy, and heat. Shifts in the patterns of man's use of energy over the last few decades and estimates of future trends are described in Figure 4–4.

Energy Storage

Early in his experience man found that the sun supplies light and heat in limited quantities both diurnally and seasonally The energy from the sun

TABLE 4–2

Summary of Quality Interchanges in Nature

SYMBOL	MEDIUM OR PROCESS	QUALITY FACTORS (PRINCIPAL)
1	Meteorological Water	Dissolved gases (CO_2, O_2, N_2), dust particles, smoke particles, bacteria, salt nuclides, dissolved vapors
2	Evaporation	Water vapor, salt nuclides
3	Evapo-transpiration Wind Pickup	Water vapor, vapors from vegetation, dust and organic particles
4	Surface Runoff	Silt, organic debris, soluble and particulate products of biodegradation of organic matter, silics, mineral residues of earth materials, bacteria, dissolved gases, soil materials
4	Infiltration to Groundwater	Dissolved minerals from surface debris and primary rocks, dissolved gases (CO_2, O_2)
5	Flood Waters	Silt and other soil materials
	Groundwaters (springs)	Mineralized water
6	Tidal Water Continental Saline Water	Increased salinity
7	River and Groundwater Discharge	Fresh water 1 and 4 plus organic debris
8	Intruded Saline Water	Increased salinity
9	Beach Erosion	Soil and vegetation
10	Evaporation	Water vapor and salt nuclides
11	Solid Residues	Biochemically unstable organic matter from life processes of animals and from death of plants and animals

From WASTE MANAGEMENT AND CONTROL. Publication No. 1400, Committee on Pollution, National Academy of Sciences, National Research Council, Washington, D.C., 1966.

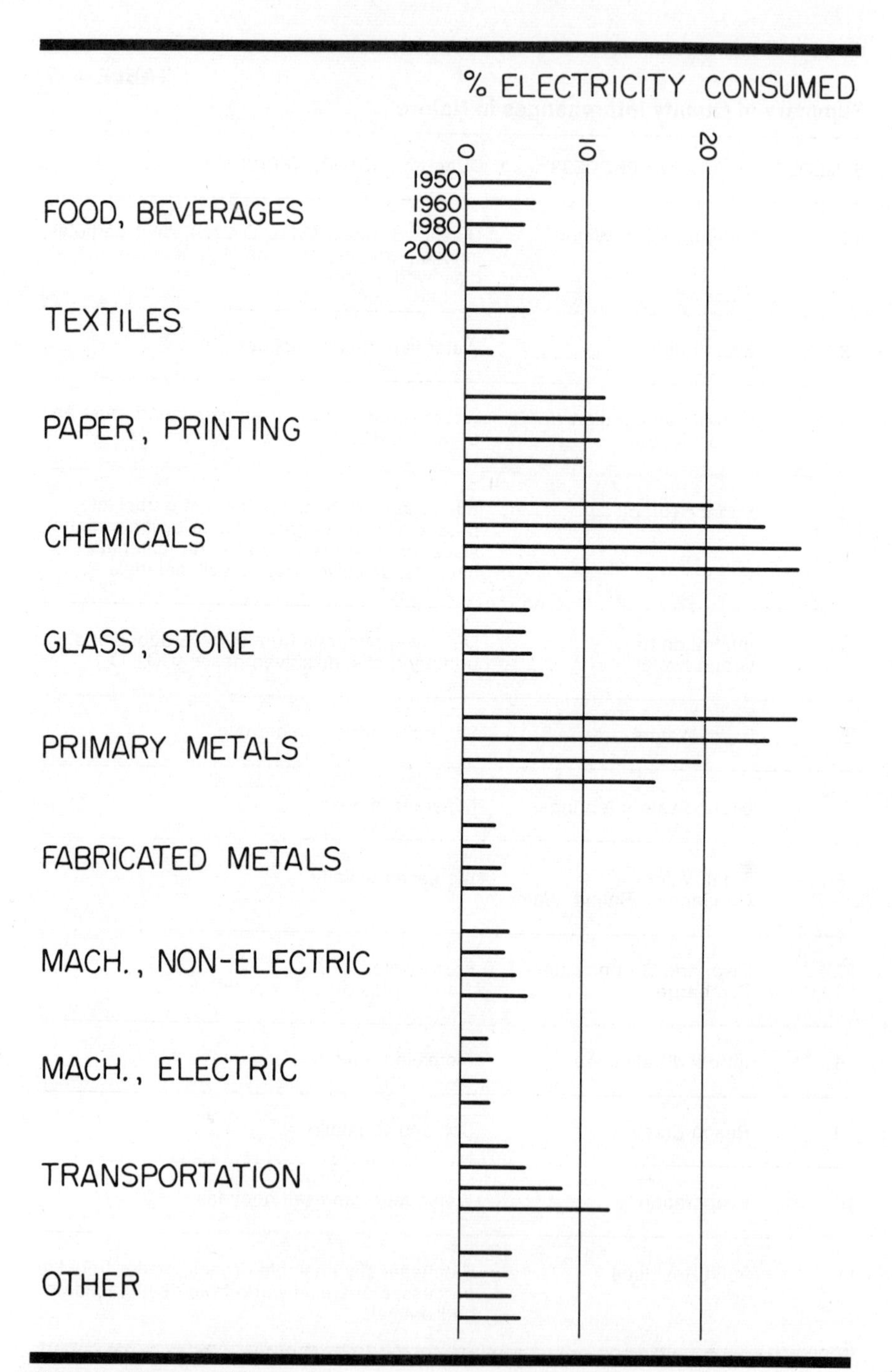

Figure 4-4. Patterns of Energy Use

From RESOURCES IN AMERICA'S FUTURE, by H. Landeberg, L. Fischman, and J. Fisher. Published for Resources for the Future, Inc. by The Johns Hopkins Press, 1963. Reprinted by permission of the publisher.

is also limited spatially in the arctic, temperate, and tropical zones. The arctic areas are deficient in terms of heat and light during much of the year. Man discovered methods of storing and transporting energy in order to make use of energy-deficient areas. Since none of the desired forms of energy (light, mechanical energy, heat, and electricity) can be stored, they had to be made, transported, and used. Forms of stored energy were required which could be converted when needed. Forms of stored energy currently used by man are water power, chemicals, fossil fuels, and nuclear fuels. Production of energy from stored fuels has particular requirements; for example, in transportation, stored fuels must have high energy content per unit weight so that they may be transported. On the other hand, to supply electricity the energy supplies must be very large, low in cost, and easily transported over very long distances. The use of electrical power for transportation requires a source, a battery or a fuel cell, which is light enough to be carried in the vehicle and which has enough capacity to drive the vehicle for long periods of time. The alternative is the use of overhead wires or underground tracks to supply electrical energy to the vehicles.

Conversion Systems

To understand the energy-waste problem, it is helpful to classify the various systems man uses to convert energy into different forms. First, there are direct natural sources, such as heat and light from the sun, which may be limited in quantity, quality, time, and space. Two-step natural systems exist by which the sun's energy is introduced into the bio-cycle to create fuel. If the fuel is obtained in the form of food, man can convert it into chemical-muscle and mechanical energy for doing work. There are also two-step systems which man uses to harness the energy available from the sun. Solar cells are very inefficient devices, just recently developed, which convert energy from the sun directly into electricity. These devices are currently being used in space exploration projects but have the disadvantage of being very costly and of requiring very large areas over which the devices must receive the sun's energy.

More conventional energy conversion systems require several man-made devices or steps to release energy stored by the natural systems. One such process would be to introduce energy from the sun into the bio-cycle that produces wood or organic matter. The wood could be harvested and introduced directly into combustion chambers or furnaces to supply heat and light. Over millions of years, products of the bio-cycle have been converted into fossil fuels such as coal, natural gas, and oil. These can be used to produce heat and light.

A further device, the engine, is required to convert the products of a combustion process into mechanical work. There are several minor energy sources which occur naturally and which man may develop for his uses, including the energy of tidal power, which drives turbines and creates mechanical or electrical energy; wind mills, which use air currents to create both electrical and mechanical energy; and natural steam sources, which can be tapped underground and harnessed to drive turbines. The major source of energy from a natural cycle is the hydrological cycle, which carries water from the oceans to the highlands, from where it returns by gravity to the sea. This water is stored behind dams and released when required through turbines to produce electrical energy.

To convert energy into usable form, man uses three devices: the combustion chamber, the engine, and the turbine. The combustion chamber is a device that can release stored energy and convert it into either heat, light, or mechanical energy. The engine takes the product of combustion chambers and converts it into mechanical energy. The turbine takes the product of either a combustion device or an engine and converts it into electricity. Thus, heat energy can be transformed to light, light energy can be transformed to electrical energy, and electrical energy can be converted to mechanical energy. In fact, any one of these four forms of energy can be converted to any other form. An examination of the applications of energy by man will reveal the various sizes and restrictions that these devices possess and the wastes that they produce.

A combustion chamber which accepts wood and fossil fuel produces ashes, which can either be discarded on land or returned to the atmosphere as particulate matter. The chamber can produce partially burned gases as a result of incomplete combustion, which are also released to the atmosphere. Incomplete burning of sulfur-containing fuels provides malodorous gases. The combustion device also creates thermal pollution. The size of various combustion devices varies from very small burners to very large power plants that can supply the energy needs of millions of people. The device that man has developed to supply mechanical work directly from fossil fuels is called a motor. This combines the combustion chamber and the engine into a single unit. Motors are used generally for transportation and produce wastes polluting mainly the air, although motor lubrication wastes pose a threat to water systems. A more recent improvement on combustion chambers is the use of uranium as fuel and the capture of energy by fission in nuclear reactors. Waste products of a reactor include radiation, radioactive materials, and heat.

Turbines are used to convert combustion energy into both electrical energy and mechanical energy, and are used to propel large naval vessels and to drive electrical turbines in mechanical plants. Turbines range in size from miniatures which fit in the palm of the hand to those found in the power houses at Grand Coulee Dam. A large turbine used to generate electrical power can produce approximately 1,000 megawatts of power. (A thousand megawatts is the equivalent of a million kilowatts, sufficient power to operate two million TV sets.)

Devices which are stationary and produce very large amounts of power present an identifiable, concentrated-point source of waste, whereas small dispersed power sources found in transportation systems[4] are highly mobile and create air pollution problems which are difficult to identify and correct.

The processes of capturing and converting energy produce their own wastes. Strip mining, drilling natural gas and oil wells, oil refining, and processing and refining of uranium ores are good examples of such processes.

Energy Source Shifts

One of the concerns in the search for energy is the possibility that sources of fuel will run out. The history of energy use suggests that man has shifted from one source to another prior to the depletion of any one source. This has been due to technological advances which create more efficient devices and use other forms of fuel. For example, Figure 4–5 indicates the decline of animal power and the increase of mechanical power in the United States during the period from 1940 to 1960.

If one were to examine the methods man has used to produce electrical power, one could see that he has shifted from wood and coal to fossil fuels and hydropower. Currently, the transition is being made to nuclear power. A similar transition can be observed in the change of propulsion units and the types of fuel used by the aircraft and space industry. For example, the first attempt to propel an aircraft was made by a man pumping a bicycle-type propeller. Next, the development of the internal combustion engine provided mechanical power from gasoline. This was followed by the jet engine and rockets for space travel. In the future, ion propulsion or even light beams appear as possibilities for major sources of transportation power. Changing the sources of fuel and the devices for converting energy causes changes in the types of wastes produced. Had we postulated back in the early 1900s that our major transportation wastes in the 1970s would be horse feces, we would have planned a grossly inadequate mechanism to control our current wastes.

The Automobile

In light of today's problems, perhaps the most significant single source of energy conversion wastes is the automobile. The automobile emits an amount of waste that can be lethal or at least extremely irritating even when greatly diluted. As long as 15 years ago it was known that the motor vehicle was the most important source of pollution in the Los Angeles basin and certainly the most difficult one to control. Meteorological predictions of the effects of auto exhaust emissions on urban atmospheres are not difficult to make. The occurrence of thermal inversions and the reactions between the chemicals in the exhaust can be shown to produce the kinds of effects that are commonly observed in the Los Angeles basin. The important point to remember is

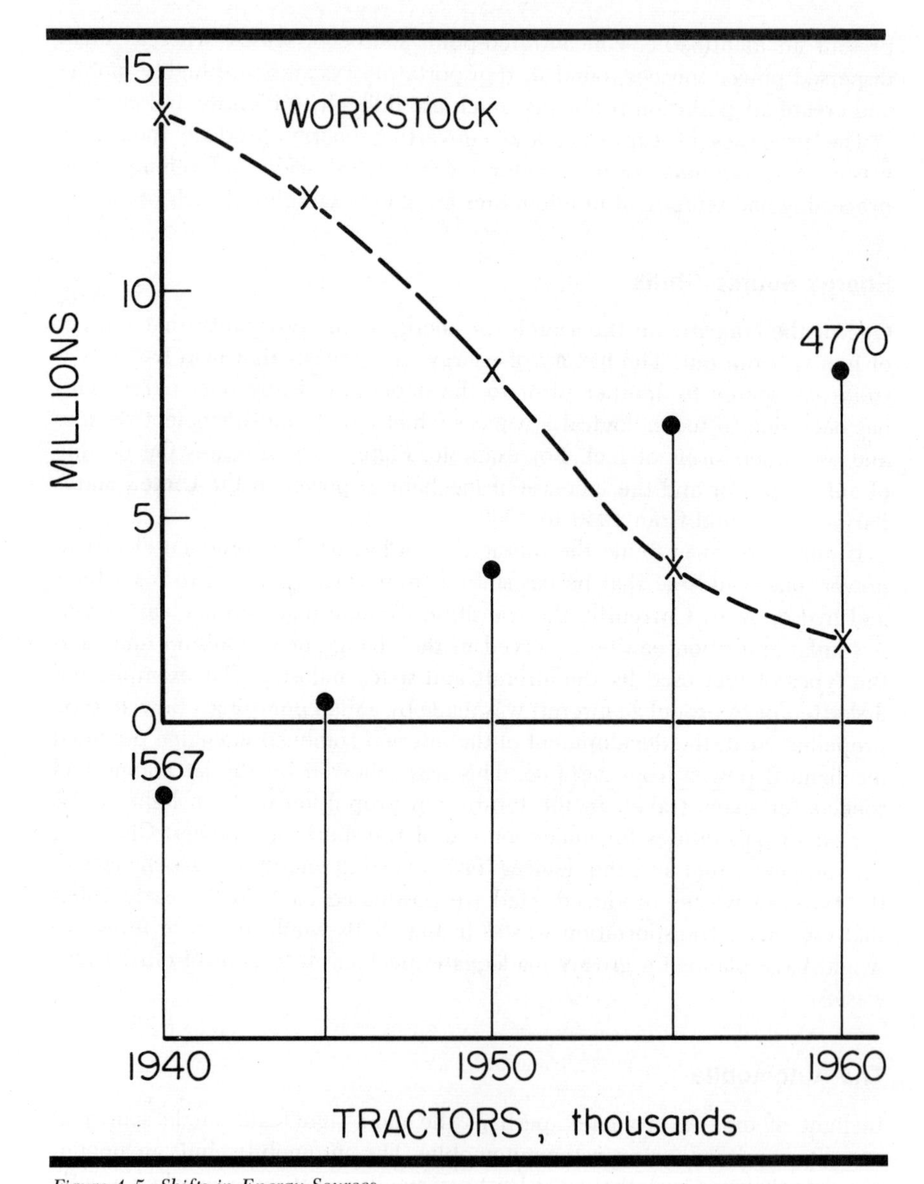

Figure 4–5. Shifts in Energy Sources

From RESOURCES IN AMERICA'S FUTURE, by H. Landeberg, L. Fischman, and J. Fisher. Published for Resources for the Future, Inc., by The Johns Hopkins Press, 1963. Reprinted by permission of the publisher.

that although they can be predicted there is no way to use meteorological information to control the effects of auto exhausts on our atmosphere short of actually preventing vehicular movement during the severest conditions. The most acceptable alternative is to modify individual vehicular transportation.

Attempts have been made during the last 10 years to persuade the automobile industry to devise exhaust control devices or engines which create fewer emissions, but results have been slow in coming. Because of the great increase in the number of vehicles on United States highways, a device removing 50 percent of the atmospheric pollutants would provide only the atmospheric conditions of 10 years ago, when the situation was already very serious. Alternatives to the use of the internal combustion engine should be sought. Perhaps the electrical automobile is one of the most feasible at this time. Of course, the use of mass transit to provide transportation for our urban population during critical hours is extremely appealing in terms of total waste emitted to the atmosphere, especially if mass transit routes are fitted to desired travel routes. Secretary of Transportation John Volpe has indicated that it may be necessary to regulate the use of automobiles in the nation's cities, unless a mass transit breakthrough is achieved. This type of control would close certain limited areas within a city to automobiles or institute fees for automobile travel in certain sections. It is more than ironic that, as long as 50 years ago, the average speed of transportation in New York City was approximately 11 miles per hour, while today the average speed in the city is 7 miles per hour despite fantastic advances in the power of automobiles. Certainly this is not progress; something must be done, and done in a bold and imaginative way.

The actual magnitude of emissions from automobiles is staggering. The average annual values for the principal pollutants (carbon monoxide, oxides of nitrogen, and the hydrocarbons) are indicated in Table 4–3.

Care must be used in judging the significance of the quantities in Table 4–3 since the tonnages alone do not reflect the effect of individual pollutants. While the effects will be discussed separately in Chapter 5, it is useful at this point to indicate the sort of reasoning that must be used in developing a hierarchy of the importance of various pollutants for a control strategy. In many instances in the past, tonnages alone have been used as a basis for control legislation and for political discussion.

One approach that could be used is developed by considering the degree to which raw automobile exhaust is diluted in the atmosphere, and the relative dilution of substances from other, stationary sources. The principal automotive emissions (carbon monoxide, oxides of nitrogen, and hydrocarbons) are diluted by perhaps 1,000 to 10,000 times in the wake of the vehicle, and are further diluted as the turbulence of the lower atmosphere takes effect.

In contrast, the emissions from stationary chimneys are diluted by 100,000 to 1 million times in most cases by the time the plume touches the ground. The result of this consideration alone suggests that vehicular emissions are somewhere around 100 times more important contributors to ground level concentrations of pollutants than are stationary sources of the same substances on a ton-for-ton basis. This consideration emphasizes the importance of automobiles as air pollutors beyond the simple tonnage consideration. As an example from Table 4–3, automobiles produce about half of the total oxides of nitrogen (NO_x), while industry and electric power generation together produce about an equal amount. The latter, however, usually emit NO_x from tall chimneys or smoke stacks so that the NO_x concentrations at ground level are due almost totally to vehicles. In locations where no automobiles exist, of course, this statement is not true, but in a city, NO_x at ground level is usually due entirely to automobiles.

If we consider the health effects of the various pollutants, we can once again develop a basis for ranking pollutants and sources. For example, carbon monoxide (CO) is perhaps 50 times less toxic than NO_2 (one particularly bad component of NO_x). Thus, although there is 6 times more CO than NO_x emitted, the NO_x may be of greater health concern than CO by a factor of 10. Of course, the type of health effect exerted by these two substances is different, so it is not possible to carry this sort of analysis very far.

These two simple examples should serve as caveats to those who would base control strategy on Table 4–3 alone. In the final analysis, many considerations must be included before deciding which emissions to control to what degree.

Carbon monoxide and the hydrocarbons are the products from incomplete burning of gasoline mixtures, while the oxides of nitrogen are derived from the chemical combination of oxygen with nitrogen in the cylinders of car engines at combustion temperatures. The concentration of gaseous pollutants in the larger cities and on crowded highways currently exceeds tolerable limits during the hours of maximum air pollution. Some of these pollutants are objectionable as emitted, while others change after emission into objectionable products. Probably the best known example is the reaction involving hydrocarbons and oxides of nitrogen in the presence of sunlight to produce the principal eye irritants contained in smog.

TABLE 4–3

National Air Pollutant Emissions, 1965 (Millions of Tons per Year)

	TOTALS	% OF TOTALS	CO	SO_x	HC	NO_x	PARTICLES
Automobiles	86	60%	66	1	12	6	1
Industry	23	17%	2	9	4	2	6
Electric Power Plants	20	14%	1	12	1	3	3
Space Heating	8	6%	2	3	1	1	1
Refuse Disposal	5	3%	1	1	1	1	1
TOTALS	142		72	26	19	13	12

From THE SOURCES OF AIR POLLUTION AND THEIR CONTROL, U.S. Public Health Service, Publication No. 1548, 1966, reprinted by permission of the publisher.

DOMESTIC WASTES

Metabolic Needs of Cities

We can interpret man's needs for materials and commodities to sustain life in cities as the "metabolic requirements" of the city. These metabolic requirements apply at home, at work, or for recreation, and include the materials needed to construct as well as to maintain and rebuild sections of the city. The cycle is, of course, not completed until the wastes or residues from the daily activities of the community have been incorporated back into the ecological system. Cities provide abundant evidence of the fact that the environment does not have an unlimited capacity to assimilate man's wastes.

Engineers and scientists who are directly concerned with the municipal problems of water supply and sewage disposal are familiar with quantities of raw material flow. Data on the inputs and outputs of materials from a hypothetical United States city with a population of 1 million are shown in Figure 4–6. Besides the air supply of the surrounding atmosphere, the largest quantity of material required in a city is water, followed by fuel

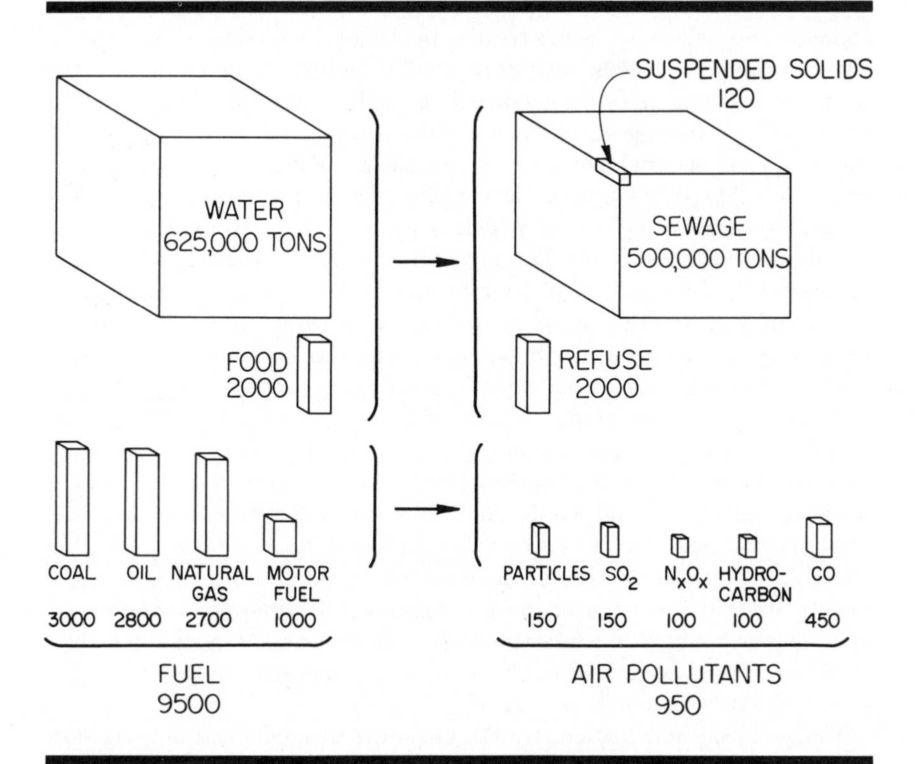

Figure 4–6. Material Flow in a City

and food. The output side shows that the waste products are sewage, solid refuse, heat, and air pollutants. Of course, there is in addition the synthesis of new cellular material. The actual quantities shown in this figure are multiples of the daily requirements of average city dwellers. About 20 percent of the water supply is diverted to lawns and other unrecoverable uses; the rest returns to the hydrological cycle as contaminated waste water through city sewers, for although the content of sewage is largely water, it contains significant quantities of suspended solids. Although the sewage produced in the hypothetical city is extremely dilute in terms of its solid concentration, it is of primary concern because of the presence of pathogenic organisms and putrescible organic material contained in it.

Perhaps the city's most pervasive nuisance is air pollution, produced chiefly by combustion of fossil fuel. Most of the particulate addition to the atmosphere derives from the use of coal as a fuel in electric power plants. Actually a large percentage of the particulate material produced daily in the city by combustion processes is removed before it reaches the atmosphere. Approximately half of the quantity of air pollutants is produced by transportation processes involving automobiles, buses, and trucks.

Several features of the solid refuse produced by a city are worthy of comment. First there has been a trend in the United States toward the production of ready-to-eat foods, such as frozen TV dinners, in which most of the food material that is brought home is actually consumed. This trend has reduced the percentage of solid refuse which is putrescible or contains unconsumed organic material, but has increased the percentage of paper and cardboard and other packaging materials in solid refuse. Approximately 50 percent of solid waste is composed of paper, 8 percent of metals, and 7 percent of ceramic and glass. Only 12 percent is garbage or putrescible organic material. One extremely significant feature of the solid refuse problem is the difficulty in handling this type of waste, although its total daily quantity is not large in comparison with sewage waste. Garbage does not flow like sewage, nor does it lend itself easily to treatment.

A careful inspection of the balance of inputs and outputs in Figure 4–6 reveals that there are approximately 130,000 tons of material not accounted for in the waste column. It is assumed in this analysis that the missing material has been converted to carbon dioxide and water, primarily through combustion processes, and thus would be introduced into the atmospheres over the cities. As was mentioned earlier there is some concern in the scientific community about the increase of carbon dioxide in the atmosphere because of man's infatuation with the internal combustion process. Although the carbon dioxide admitted to our atmosphere is not a noticeable pollutant, it may well have the most significant effects.

The problems and particularly the quantities of inputs and outputs vary widely from city to city, and are managed within the cities with varying degrees of success. As Wolman[1] has pointed out, it is ironic that New York

City, which contains a great concentration of managerial talent, should be running short of water while millions of gallons of fresh water flow past the city to the sea every day. While for several summers New Yorkers watched their reservoirs empty, Californians were busy building aqueducts to carry water over 400 miles from the Sacramento River to the southern part of the state. In other regions of the country, water shortages occur as a result of delayed action, failures of management, or political maneuvering.

Before leaving the discussion of domestic waste, several other problems should be introduced at this time which will be discussed in greater detail later. The first of these is the disposal of solid wastes, such as car bodies and construction residues, which accumulate from the growth of a city. Another problem arises from the paving of vast areas of the city and consequent elimination of the natural absorption of water, creating very large, rapid releases of water after each rainstorm. Rainwater usually enters sewer systems, further diluting the waste and creating vast problems in treating and disposing of a very dilute waste stream. In addition, the initial rains after a dry spell can carry significant loads of road oils, dirt, fertilizers, and other ground accumulations.

Evolution of the Pollution Problem

The nature of the individual water, land, and air pollution problems in all cities has gone through periodic cycles. In the specific case of water pollution, the first and foremost problem associated with water-borne waste was the problem of disease control. Pathogenic organisms carried by sewage have immediate impact on man's well-being and, in fact, bear directly on his survival in communities. The chlorination practices instigated in this country and throughout the world within the last 50 years have diminished this problem to the point where very few public waters are considered unpotable for health reasons. No sooner had this problem diminished in severity than man began to realize that the increasing strength of his sewage waste in terms of the putrescible material contained in them was a significant problem. Engineers refer to this tendency of organic wastes to degrade naturally as Biological Oxygen Demand (BOD), or the amount of oxygen a given waste will require to degrade. Man has built tremendous engineering structures to collect sewage and accelerate the natural degradation process before discharge into the receiving streams. The efficiencies of these plants are rated largely by the percentage of BOD reduction. For instance, in the city of Seattle, Metro's West Point treatment plant provides a BOD reduction of approximately 35 percent by removing solid and settleable material from the sewage, whereas Metro's Renton plant provides additional treatment in the form of biological oxidation of soluble wastes and removes approximately 90 percent of the waste's BOD. Treatment facilities designed to remove oxygen-consuming organic materials from waste do not remove fertilizing elements, toxic materials, or heat from sewage streams. One might imagine that the treatment of these elements in waste streams may become the next crisis in handling water-borne wastes.

Man's handling of his air pollution problem has gone through similar crises, beginning with the first recognition of the problems connected with emission of large settleable particulate matter from smoke stacks. After controlling this problem, man became aware of the undesirable effects of noxious odors, as well as visibility reduction caused by the smaller particles which could not be removed in first-stage treatment processes. Future efforts will be concerned with the emission of invisible gases, such as carbon dioxide, which may seriously interfere with the planetary heat budget. Similarly, solid waste handling was first concerned with disposing of putrescible, degradable organic material, and then with handling refuse largely composed of paper and car bodies. One can imagine problems in the future deriving from handling increasingly large quantities of plastic materials.

Development and control of any one of these cycles must be handled with a perspective embracing all environmental resources. For instance, the improvement of air quality as a result of scrubbing the emissions from smoke stacks would not be acceptable if the water used in the process were then discharged to lakes or estuaries to pollute their water. Similarly, solving the solid waste problem may complicate water and air pollution problems. Most noticeable in this respect is the process of incineration, which is used widely by many cities to dispose of solid wastes. When poorly managed, incineration can be a major contributor to air pollution.

The problems of material flow in cities are actually a reflection of the overall phenomenon of urbanization, which itself represents a new step in human social development. In 1960, for example, nearly 52 million Americans lived in 16 urbanized areas, and 53 percent of our nation's population was concentrated on only 0.7 percent of the nation's land. These agglomerations of urban population involve a degree of human contact and social complexity that man has not faced previously.

The most rapid rate of increase of urbanization occurred between 1950 and 1960; were that rate to continue, by the year 1990 more than half of the entire world's population would live in cities with populations larger than 100,000.

The process of urbanization is much more than the growth of cities. Cities can grow without urbanization if the rural population increases at an equal or greater rate. In the most advanced countries, urban populations are still increasing but their relation to the overall population tends to remain constant or to decrease. In other words, the transfer from a spread-out pattern of human settlement to one of concentration in restricted urban centers is a change that does have a beginning and an end, but the growth of cities itself has no inherent limit. The reader is referred to Kingsley Davis' excellent article for a thought-provoking discussion of the mechanisms of urbanization.[2]

AGRICULTURAL WASTES

The capture, growth, processing, and consumption of organic material to satisfy man's needs for food, shelter, and mobility constitute a major human activity. It has been previously mentioned that all biological systems require air, water, and food, and unless man elects to capture natural supplies for his needs and allow these to propagate naturally, he must alter the environment so that accelerated biological crops can grow. A major conceptual distinction exists between capturing natural growth and managing the environment to maximize the production of a controlled growth. Included in the category of "biological materials" are grain, fruit, other crops, meat, fish, lumber, and other forest products. The various processes involved in agricultural and related industries are shown schematically in Figure 4–7.

Capture of Crops

Major biological stocks which are captured by man include animals, fish, and forest products. Since man has learned that management and conservation of natural stocks can yield sustained yields over periods of time, a trend toward farming rather than direct capture has occurred. The major wastes which occur in the capture process derive from partial use, although the effect of imbalances in nature from man's selective harvesting of one species must also be considered. Capturing seals for their skins, capturing birds for plumage, and harvesting trees for lumber all create residual wastes which lead to disposal problems.

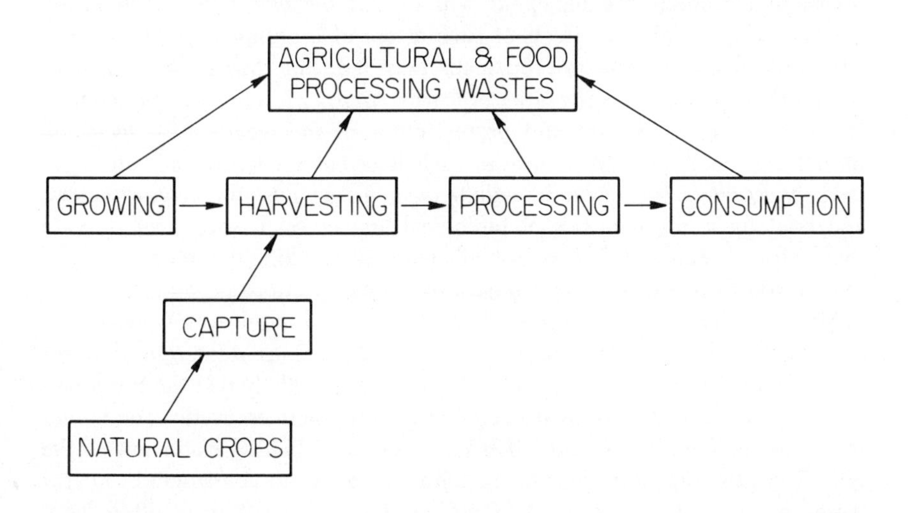

Figure 4–7. Agricultural Processes

Growth of Crops

The farming or management of crops, trees, animals, or fish is accomplished by altering the media in which the crops grow to maximize production. Steps may be taken to supply water and nutrients for land crops, to modify the climate, and to reduce the number of species that will compete with the desired crop. Each of these activities alters the environment directly and in turn creates waste which must be returned to the environment. Activities related to construction of farms are particularly important, for they often involve substantial modifications of existing environments. Even harvesting, the ultimate goal of farming, is not completely efficient and leaves residual wastes which must be returned to the environment in some manner.

There is a limited amount of land that can support crops without some type of land conditioning. Natural growth must be removed prior to planting and, of course, adds wates to the environment. There has been a shift in emphasis in this country from forest land to crop land and back again, as man has changed his demands for agricultural products. For example, many of the old cotton fields of the South are being planted with pine trees for use as paper pulp. Changes in crop patterns are influenced by soil erosion and even by changes in local weather patterns.

Significant modifications of the environment needed for growing crops often involve the control of weather. In the states of Washington, California, and Florida, smudge pots are often used to protect crops from frost damage. Even under ideal conditions, using this type of weather control often creates extremely adverse atmospheric conditions for nearby residents and hazardous conditions for farm employees. The use of wind screens for crop protection is another modification which is more unsightly than hazardous. The artificial production of rain by seeding clouds with crystals of silver iodide is attempted frequently, although the effects of such attempts are unpredictable at best. Attempts to create rain on crop lands may fail, but they may manage to trigger downpours on nearby cities or towns. Another form of weather control is the provision of shelter and protection from the weather for livestock in artificial enclosures. Wastes from such livestock enclosures are often extremely concentrated, make immediate oxygen demands on receiving waters, and contain large quantities of biological agents. Such structures provide shelter for animals and increase the efficiency of handling livestock, but deny the natural dispersal of animal wastes over large sections of land.

Much of the land that is used to grow our food requires only water for successful production. Man has been aware of the important role of water in producing his food almost from his beginning. It is likely that the Egyptians were the first to make extensive use of irrigation practices. Pictures on monuments dating back to around 2,000 B.C. show men bailing water out of the Nile River to pour onto their fields. Later, canals were constructed to carry water to more distant fields. In China and Italy, canals were often large enough to accommodate boats and were used both as a source of water for irrigation and as a means of transportation. Eventually machines were developed for lifting irrigation water into the fields. Many of these machines were crude and inefficient by modern standards, but some are still in use in China, India, and Egypt.

In addition to water from streams, early irrigators used natural lakes, ponds, and wells as sources of irrigation water. The next logical step was the building of small dams to conserve water for use during dry seasons. Low dams or weirs have been common during much of the history of irrigation; storage wells and tanks are still common today in India.

The use of irrigation water in the United States was greatly accelerated by the Homestead Act of 1902, which was designed to induce families to settle the dry lands of the 17 western states. In 1956, over 36 million acres, or approximately 2 percent of the land in the United States, was under irrigation. Over 90 percent of the irrigated land is in the western regions of the country. A frequent problem with land in need of irrigation is the accumulation of salts in the upper layers of the soil. Irrigation water must remove these salts before the water can be used for the primary purpose of encouraging plant growth. As a result, the process of irrigation tends to mineralize the water that runs over the farm land. The problem is further complicated by the fact that the water must often be impounded in very hot regions, where large quantities of water are lost through evaporation. This process increases the salinity of the return flow even more. The overall result is an increase in both the quantity of water required for irrigation and an increase in the dissolved salts in returning waters. Some water which is used for agricultural purposes, of course, seeps deeply into the ground and is purified before it returns to the hydrological cycle. The part of irrigation water that runs off agricultural land and returns directly to the stream is one source of quality degradation. A continuing degradation of water quality can result from long periods of poor irrigation practice.

Irrigation is used not only in arid regions. It is becoming a widespread practice in wetter parts of the United States to apply supplemental irrigation water at critical stages of plant growth. It has been predicted that the volume of water withdrawn from the manageable resources for agriculture in the future will increase, although not as dramatically as will withdrawals for industry.

The actual quantities of water withdrawn for agricultural use have increased steadily from 1949, when approximately 100 million acre-feet were used, to current estimates of approximately 170 million acre-feet. The actual quantity of irrigation water needed varies considerably among the 17 western states, which are the principal users. The demands vary from 2.2 acre-feet per acre per year in the Rio Grande Valley to 6.1 acre-feet per acre per year in Idaho. These rates depend not only on the nature of the crop irrigated, but also on the local cost of water, the type of soil, and the local climatic conditions. Application rates in humid climates vary from 1.25 to 5 acre-feet per acre per year.

As much as 20 to 60 percent of the water applied to irrigation fields may return rapidly to the river or stream from which it was taken. This water is referred to as irrigation return flow. Not all of the remaining percentage is actually used by the crops, as significant quantities are lost by evaporation from irrigation canals, seepage, or uptake from plants growing along the irrigation works. Estimates of the total loss are around 35 percent of the total reservoir capacity available for irrigation. These evaporative losses concentrate salts and other soil minerals that are picked up by water in transit through irrigation fields. It has been estimated that for irrigation in the western states, approximately one third of the water diverted returns more or less immediately in the irrigation return flow. From this one can predict that the salt concentration in the return water will be multiplied by a factor of 3. Actual chemical measurements in the return waters of the western irrigation fields reveal that this estimate is perhaps low and may be due to the fact that return flows were less than one third or that the water was cycled more than once.

Perhaps the most significant effect of changing the chemical composition of agricultural return waters is the great increase in the hardness of the water. The total dissolved solids in the water are also greatly increased, and other physical properties of the water such as its temperature, its turbidity, and even its color may be measurably affected. The increase in concentration of nutrient minerals such as nitrogen and phosphorus is of additional concern, because it can lead to nuisance algal blooms and eventually to the development of unpleasant tastes and odors. Insecticides and herbicides also can seriously degrade the quality of irrigation return flow. This depends on the quantity and the method of application. Bacteria or other biological contaminates are not significant factors in the return waters from agricultural practice unless sewage is used for irrigation.

Not all of the chemicals picked up by irrigation waters are present naturally in the soil. Many chemicals are added as fertilizers containing nitrogen, phosphorus, and potassium. The quantity applied to each acre of farmland depends on the type of soil and the particular crop, but may amount to several hundred pounds per acre. Agricultural return waters contribute significantly to silt and other sediments which are transported by rapidly flowing waters back into the natural reservoir. The result is the accumulation of sediments that may seriously interfere with the natural productivity of open waters.

Quality degradation in agricultural return waters is of major importance since agriculture uses extremely large portions of water resources. This problem poses two major challenges. First, the management of farm practice needs careful attention so that the quality of irrigation water does not degrade to the point where continuing benefits from irrigation can no longer be derived. There is a tremendous educational and training challenge in seeing that individual farmers understand the relation of their return water quality to the overall quality of the surrounding environment. Care must be exercised in applying and distributing fertilizers and pesticides; large quantities of animal waste must be kept from contact with the return water. The second challenge deals with the management of agricultural return waters on a regional basis. The same reservoir which supplies irrigation water may have other uses, and the resource must be managed under policies which clearly appreciate the nature and magnitude of each use. These management policies must recognize the potential conflict that exists between environmental factors and agricultural practices.

Harvest of Crops

Once a crop is grown it must be harvested, and wastes from this process may be particularly troublesome in that they often occur at one particular season of the year. Wastes from harvesting operations include parts of the crop which are undesirable, such as leaves, roots, vegetable coverings, stalks from vegetables such as corn or grain, leaves and stumps from tree farming, poultry feathers, and even unsightly creations such as access or timber roads in woodland areas. As this material accumulates at the harvest site, a decision is needed as to the ultimate disposition of the solid waste and the organisms associated with the crop. Should these degrading materials remain and be allowed to decay naturally, or should they be collected and burned? Should they be thrown in the water, or should they be trucked away from the harvest site and buried? The problems of cost, disease control, ease of handling, availability of cover, odor, and aesthetics influence decisions as to how these residues will be returned to the environment.

Crop Processing

Processing harvested crops prior to sale and consumption often creates the most concentrated waste in the entire agricultural and food-processing industry. Processing generally involves many separate steps: preparation of the crop, preservation of the desirable portion of the crop, and packaging of the crop for distribution. Preparation involves such processes as shucking corn, gutting and skinning animals, removing peas from pods, juicing of foods such as sugar beets, and separating soil from the crop. To prepare cellulose wood pulp for the paper industry, wood must be debarked and then treated with strong chemical solutions to remove the lignin from the woody tissue so that the cellulose may be further processed. These wastes are often concentrated and toxic, and include strong acidic or alkaline solutions used to modify, isolate, wash, bleach, or otherwise convert the crop prior to further processing. The crop, once prepared, may be distributed fresh or it may be chemically processed; some foods are canned, some dried, and some frozen. Packaging crops creates solid wastes in the form of spillage or trimmings from the packing operations. Disposal of the container and waste from the finished product is passed on as consumer waste.

A large section of what is commonly referred to as pollution technology stems from processing agricultural products. These wastes are often referred to as food processing wastes, or, in cases involving large quantities of raw materials and drastic chemical reactions such as in pulp and paper operations, simply industrial wastes. Many examples of industrial wastes will be discussed later in terms of their specific effects on the environment.

Consumption

It is a mistake to think of the consumption unit of the agricultural process referred to in Figure 4–7 as the ultimate end of agricultural products. Rather, agricultural products are used by man and by his industries, then returned in an altered form to the environment; they are not literally consumed. The use of agricultural products and the subsequent formation of sewage or solid refuse was covered in a previous section. Industries are large consumers of agricultural products; for examples, one might cite the brewery industry, the wood and paper industry, and the dairy industry. Some of the waste-streams from these users of agricultural products are highly concentrated, generate heat and noise, and introduce chemical pollutants to water and air resources.

It is important to realize that reusing any waste products from growing, harvesting, processing, and consumption can minimize the final or total waste problem in the environment, and at the same time may help to maximize agricultural production. The Simplot organization in Idaho provides an outstanding example of diversity in agriculture which makes maximum use of by-products. In this organization, farmlands are used to grow barley, alfalfa, seed potatoes, and potatoes for wholesale. Seed potatoes from the farms are grown by contract growers, and the fresh potatoes are shipped to processing plants where they are converted to packaged potato products. Barley and alfalfa products from the farms are used to feed sheep and cattle. The waste products from potato processing provide nourishment in cattle feedlots; whole animal wastes from the feedlots are used in turn as fertilizer on the farmlands. A more recent addition to the system has been the use of potato processing residues as chicken feed. The Simplot organization also is an excellent example of interplay between agriculture and other industries in its use of phosphate mines to produce fertilizer, and saw mills to produce lumber and wooden boxes needed for shipping.

INDUSTRIAL WASTES

Industry is one of the major uses of water in the United States, and like agricultural and domestic users causes an increase in the concentration of dissolved and suspended material in the water. The similarity ceases at this general level, however, because of the tremendous variety of industrial operations, and the corresponding variety of wastes in their effluent waters. Many industries discharge process waters which contain chemical compounds not found in natural water. Many contain toxic metal ions and exotic organic and inorganic chemicals, as well as compounds which are refractory or stable in nature. Industrial waste streams frequently have high temperatures and are turbid, discolored, acid, or alkaline in degrees that are very much unlike other wastes associated with man's use of water.

Projections of water use for the future clearly indicate the growing importance of industry as a consumer of water. Among the various industrial uses, steam power is the fastest growing. Its general effect is to reduce the overall amount of water and thereby increase the effects of wastes discharged by other users. Of the water withdrawn by industry, approximately a third may be returned with the chemical salt content doubled, another third may be contaminated with organic and inorganic solids from industrial process, and the other third is consumed by incorporation into a product or by evaporative loss to the atmosphere. Projections indicate that the pollution load on water resources through industrial consumptive depletion and return of effluent waters is growing by approximately 30 percent per decade.

Because the word "industry" refers to many economic activities, the effects of industrial pollution cannot be judged on a per capita basis as domestic wastes. Rather, they must be judged individually in terms of the amount of water used and the particular nature of the effluent. There are, of course, great numbers of small industrial operations whose waste products may create serious local pollution conditions; however, most of the total industrial use may be characterized in half a dozen or more general categories as shown in Table 4–4.

Just as total water use among the industries varies, so do the water requirements in gallons per ton of product. The largest users of water in these terms would be the chemical industry, sulfite pulping operations, and the textile industry, particularly operations involved in producing rayon. It should also be pointed out that water use by industry varies considerably with the water quality objectives enforced in any local situation. Thus, pulp mills in some regions of the country may be designed to use less water than similar mills in regions where water is more plentiful or pollution control is more lax. It is extremely difficult to classify types of industrial waste, although such classifications may be made on the basis of whether the waste contains principally minerals or organic materials. A simple classification of this type is shown in Table 4–5.

Virtually all industrial operations contribute to atmospheric pollution, some in greater degree than others. This is because most industrial operations emit particulate material which reduces visibility. Foul odors can be caused by such operations as pulp and paper manufacture, slaughter-houses, food processing activities, the chemical industry generally, and tanneries. Industrial operations which emit reactive gases are another source of atmospheric pollution: copper refining, the production of electricity from coal and oil with relatively high sulfur contents, painting and metal cleaning operations, and the transportation industry are particularly noteworthy. It must be emphasized here that the emissions from the transportation industry, particularly the automobile, have an extraordinary if not overwhelming impact on the atmosphere. In many western cities with freeway systems, it has been estimated that from 65 to 85 percent of the total atmospheric pollution load derives from automobiles, trucks, and buses.

A partial summary of industries producing waste, the origin of their major wastes, and principal characteristics is presented in Table 4–6.

In the rest of this section we will focus our attention on several selected industrial operations for a closer examination of the processes that produce the waste and the characteristics of the wastes themselves.

TABLE 4–4

Annual Water Use by Industry, 1968

TYPE OF INDUSTRY	ANNUAL USE IN GALLONS	PERCENTAGE OF TOTAL INDUSTRIAL USE
Metal Manufacture	7.8×10^{12}	22
Chemicals & Allied Products	9.4×10^{12}	26
Paper Industry	6.5×10^{12}	18
Petroleum & Coal Products	9.4×10^{12}	26
Food Industry	1.3×10^{12}	4
Textile Products	0.3×10^{12}	Less than 1
Rubber & Plastics	0.27×10^{12}	Less than 1
TOTAL	35.7×10^{12}	98

From U.S. Bureau of Census, Census of Manufacturers, Vol. 1, Summary and Subject Statistics, Chapter 7. Water Use in Manufacturing (1968).

TABLE 4–5

Types of Industrial Wastes

CHIEFLY MINERAL CONTAMINANTS	CHIEFLY ORGANIC CONTAMINANTS
Brine wastes	Petroleum refining
Mine drainage	Gasoline station and Garage wastes
Electro-plating wastes	Organic chemical manufacture
Water softening	Paint and varnish wastes
Boiler blowdowns	Chemical Plant wastes
Inorganic chemical wastes	Wood distillation wastes
Photographic wastes	Pharmaceutical wastes
Cooling water	Laundry wastes
	Textile manufacturing wastes
	Food processing wastes

TABLE 4–6

Summary of Industrial Waste: Its Origin, Character, and Treatment

Industries Producing Wastes	Origin of Major Wastes	Major Characteristics (Water)	Major Characteristics (Air)
Food and Drugs			
Canned goods	Trimming, culling, juicing, and blanching of fruits and vegetables	High in suspended solids, colloidal and dissolved organic matter	
Dairy products	Dilutions of whole milk, separated milk, buttermilk, and whey	High in dissolved organic matter, mainly protein, fat, and lactose	
Brewed and distilled beverages	Steeping and pressing of grain, residue from distillation of alcohol, condensate from stillage evaporation	High in dissolved organic solids, containing nitrogen and fermented starches or their products	Odiferous organic compounds
Meat and poultry products	Stockyards, slaughtering of animals, rendering of bones and fats, residues in condensates, grease and wash water, picking of chickens	High in dissolved and suspended organic matter, blood, other proteins, and fats	Odiferous, possibly high in airborne bacteria
Beet Sugar	Transfer, screening and juicing waters, draining from lime sludge, condensates after evaporator, juice, extracted sugar	High in dissolved and suspended organic matter, containing sugar and protein	
Pharmaceutical	Mycelium, spent filtrate, and wash waters	High in suspended and dissolved organic matter, including vitamins	
Yeast	Residue from yeast filtration	High in solids (mainly organic) and BOD	
Pickles	Lime water; brine, alum and turmeric, syrup, seeds, and pieces of cucumber	Variable pH, high suspended solids, color, and organic matter	
Coffee	Pulping and fermenting of coffee bean roasting	High BOD and suspended solids	Odiferous gases
Fish	Rejects from centrifuge, pressed fish, evaporator and other wash water wastes solids	Very high BOD, total organic content	Odiferous

TABLE 4–6 (Cont.)

Industries Producing Wastes	Origin of Major Wastes	Major Characteristics (Water)	(Air)
Rice	Soaking, cooking, and washing of rice	High in BOD, total and suspended solids (mainly starch)	
Soft drinks	Bottle washing, floor and equipment cleaning, syrup-storage-tank drains	High pH, suspended solids, and BOD	
		Apparel	
Textiles	Cooking of fibers, desizing of fabric	Highly alkaline, colored, high BOD and temperature, high suspended solids	
Leather goods	Unhairing, soaking, deliming and bating of hides, evaporation of chemicals	High total solids, hardness, salt sulfides, chromium, pH, precipitated lime and BOD	Odiferous gases
Laundry trades	Washing of fabrics	High turbidity, alkalinity, and organic solids	Lint and dust from ventilating air
		Chemicals	
Acids	Dilute wash waters, many varied dilute acids, some volatilization	Low pH, low organic contents	Acid anhydrides, often gaseous: SO_2, NO_2, etc.
Detergents	Washing and purifying soaps and detergents	High in BOD and saponified soaps	
Cornstarch	Evaporator condensate, syrup from final washes, wastes from "bottling up" process	High BOD and dissolved organic matter: mainly starch and related material	
Explosives	Washing TNT and guncotton for purification, washing and pickling of cartridges	TNT, colored, acid, odorous; contains organic acids and alcohol from powder and cotton, metal, acid, oils, and soaps	

Insecticides	Washing and purification products such as 2, 4–D, and DDT	High organic matter, benzene ring structure, toxic to bacteria and fish, acid	Insecticides' *use* puts more in air
Phosphate and phosphorus	Washing, screening, floating rock, condenser bleed-off from phosphate reduction plant	Clays, slimes and tall oils, low pH, high suspended solids, phosphorus, silica, and fluoride	
Formaldehyde	Residues from manufacturing synthetic resins, and from dyeing synthetic fibers	Normally has high BOD and HCHO, toxic to bacteria in high concentrations	Reactive odiferous gases
		Materials	
Pulp and paper	Cooking, refining, washing of fibers, screening of paper pulp	High or low pH, colored; high suspended, colloidal, and dissolved solids; inorganic fillers	Odiferous mercaptan vapors, particulate matter high in water-soluble salts
Photographic products	Spent solutions of developer and fixer	Alkaline, contains various organic and inorganic reducing agents	
Steel	Coking of coal, washing of blast furnace flue gases, and pickling of steel	Low pH, acids, cyanogen, phenol, ore, coke, limestone, alkali, oils, mill scale, and fine suspended solids	Colored particulate matter as a smoke
Metal-plated products	Stripping of oxides, cleaning and plating of metals	Acid, metals, toxic, low volume, mainly mineral matter	
Iron-foundry products	Wasting of used sand by hydraulic discharge, smoke	High suspended solids, mainly sand; some clay and coal	Same problems as steel industries
Oil	Drilling muds, salt, oil, and some natural gas, acid sludges and miscellaneous oils from refining, evaporation, incineration	High dissolved salts from field, high BOD, odor, phenol, and sulfur compounds from refinery	Reactive organic vapors, smoke, odor
Rubber	Washing of latex, coagulated rubber, exuded impurities from crude rubber	High BOD and odor, high suspended solids, variable pH, high chlorides	
Glass	Polishing and cleaning of glass	Red color, alkaline non-settleable suspended solids	

TABLE 4–6 (Cont.)

Industries Producing Wastes	Origin of Major Wastes	Major Characteristics (Water)	(Air)
Naval stores	Washing of stumps, drop solution, solvent recovery, and oil recovery water	Acid, high BOD	
		Energy	
Steam power	Cooling water, boiler blowdown, coal drainage, smoke	Hot, high volume, high inorganic and dissolved solids	Flyash, soot, sulfur oxides, and nitrogen oxides
Coal processing	Cleaning and classification of coal, leaching of sulfur strata with water, de-sulfuring	High suspended solids, mainly coal; low pH, high H_2SO_4 and $FeSO_4$	Sulfur oxides, odor
Nuclear power and radioactive materials	Processing ores, laundering of contaminated clothes, research-lab wastes, processing of fuel, power-plant cooling waters, gas waste, leaks	Radioactive elements; can be very acid and "hot"	Radioactive gases
		Transportation	
Automobiles, trucks, etc.	Combustion of fuel, tire wear, evaporation of fuel	Dustfall into reservoirs	Reactive and/or toxic gases, NO, NO_2, CO, hydrocarbons; particulate matter; soot, lead compounds

From LIQUID WASTE OF INDUSTRY, Theories, Practices and Treatment, by Nelson L. Nemerow. Copyright © 1971 by Addison-Wesley, Reading, Mass. Reprinted by permission of the publisher.

Pulp Mill Wastes

The Pacific Northwest accounts for approximately 16 percent of all United States pulping operations. The various pulping processes, Kraft, sulfite, and ground wood operations, amount to 12, 40, and 30 percent of the United States market, respectively. Some idea of the financial magnitude of pulping operations in the Pacific Northwest is given by the fact that 20 percent of the total annual manufacturing and agricultural sales in the state of Washington comes from pulping processes, compared with the 22 percent from the entire aerospace industry.

An appreciation of the waste problem associated with obtaining pulp from wood may be obtained from a realization that the desirable commodity in wood is the cellulose and that cellulose comprises only about half the tree. Subsidiary pulping operations such as sawmills waste about 20 percent of each log as sawdust and shavings. These wastes are also fed into pulping operations from which cellulose is derived. In fact, as much as 70 percent of the raw materials fed into pulp mills is obtained from the wood waste products of these subsidiary operations, indicating a somewhat economical use of forest products by the forest industry. The fact still remains, however, that of any wood which is sent into a pulp mill, approximately half must appear as waste in one form or another, at least until lignins are put to some useful industrial purpose.

There is a tendency in the pulping industry to increase the size of mills. Older pulp mills had capacities of 20 to 500 tons per day, whereas mills currently being constructed may produce 1000 tons per day. Projections for the future indicate sizes up to 1500 tons per day. Whereas this trend may in fact decrease regional pollution problems, conditions in the immediate vicinity of larger plants may worsen. There is a distinctly different problem connected with Kraft mill pulping than with sulfite mill pulping: the Kraft process generally leads to air pollution problems, whereas the sulfite mill constitutes a substantial water pollution problem. These processes will be taken up individually.

There are 22 pulp and paper plants in the state of Washington; these plants discharge a total of nearly 500 million gallons per day of waste water. The average volume of waste water discharge from each plant is about 24 million gallons per day. Of these plants, four are sulfite mills, four are Kraft mills, five are ground wood mills, and one is a soda base operation. Those remaining are paper or tissue mills.

The major process objective in any operation is to remove the structural element of the tree, namely the lignin, leaving the cellulose fiber free to be filtered out, bleached, washed, and subsequently processed into pulp. The Kraft and sulfite processes differ primarily in the types of chemicals that are used to dissolve the lignin. In the Kraft process, very alkaline solutions of sodium sulfate are used, whereas in the sulfite process extremely acid solutions of metal bisulfite are employed. Wood chips are digested in the various chemicals for several hours in large pressure vessels, known as digesters, at pressures which may reach hundreds of pounds per square inch. After cooking, the contents are dumped into a blowtank where the liquor containing the dissolved lignin is drained and the pulp may be further processed for washing, bleaching and pressing. A typical process flow diagram for a sulfite mill operation is shown in Figure 4–8.

Pulping plant effluents have certain effects on natural streams into which they are discharged. As has been mentioned, over half of the organic matter present in wood is transformed into waste products, and if these are discharged without treatment or recovery into natural receiving waters, adverse effects may result. For example, as little as 50 parts per million of Kraft effluent has been found to depress the activity of oysters and decrease their gill pumping rates. Larger concentrations of Kraft wastes may cause a receiving stream to lose its oxygen content, producing septic substances such as

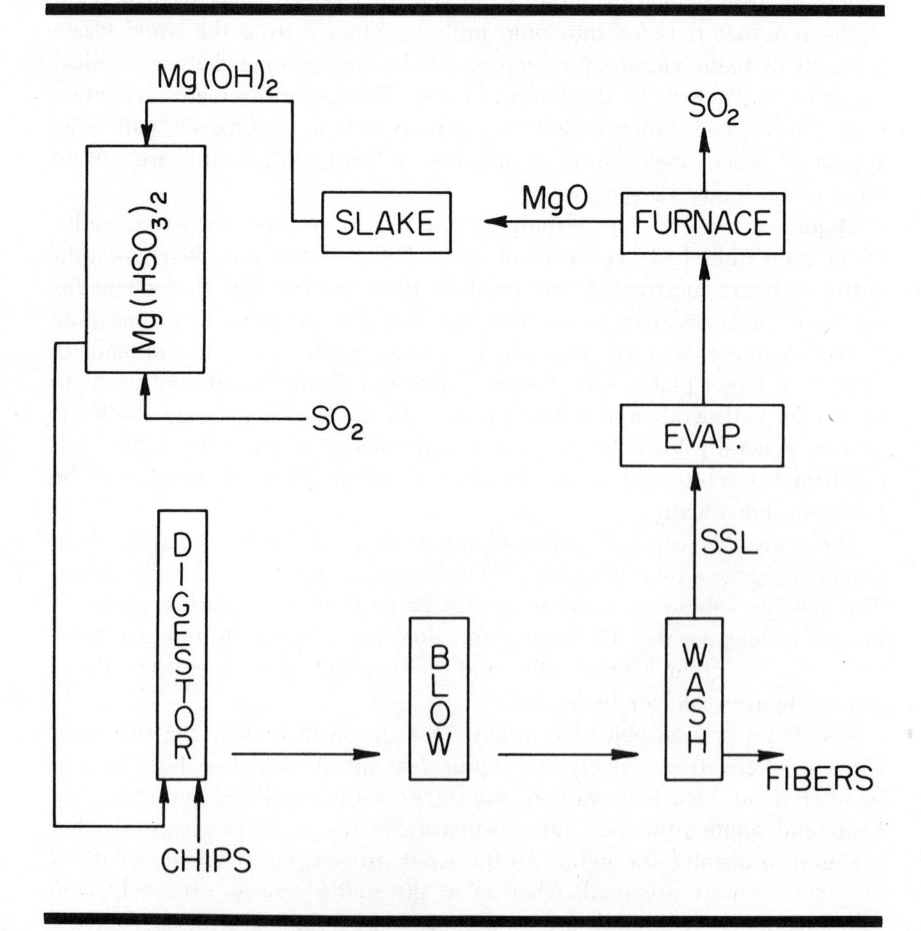

Figure 4–8. Sulfite Pulping Process Flow Diagram

hydrogen sulfide, which is toxic to fish and has an unpleasant odor. Low oxygen content can lead to uncontrolled growth of fungi, causing a body of water to become esthetically undesirable in both appearance and odor.

As far as water pollution is concerned, Kraft pulp wastes produce an effluent of higher quality than do sulfite mills. Sulfite wastes contain larger quantities of biodegradable organic material which can cause more serious oxygen depletions in the receiving water. The Kraft process produces noticeable air pollution by releasing extremely noxious gases, such as sulfides and sulfur-containing gases such as the mercaptans, in relatively large quantities. The odor near Kraft mills is extremely foul and smells like rotten eggs. The principal gas emitted from sulfide mills is sulfur dioxide, which accounts for the acrid smell around these mills. Sulfite operations consume approximately twice as much water as Kraft mills, produce twice the amount of dissolved chemicals in the waste, have higher concentrations of total solids, and have greater BOD values than Kraft mills. In addition the sulfite effluent is extremely acid whereas the Kraft effluent is usually alkaline. However, sulfite effluents generally contain fewer suspended solids than Kraft effluents.

Petroleum Refinery Wastes

The petroleum industry is extremely diversified and quite complex. The various operations conducted at petroleum refineries include storing bulk oil, desalting crude oil, crude oil fractionation, thermal cracking, catalytic cracking, polymerization, solvent refining, dewaxing, de-asphalting, wax manufacture, and hydrogen manufacture. Any of these processes results in the accumulation of waste which must be discharged.

Several problems are generally connected with petroleum refinery wastes. The temperature of effluents may reach 70° to 100°F. The refining operations generally introduce considerable quantities of soluble hydrocarbons and phenols into waste waters, causing odors and oxygen consumption in receiving waters. In addition there is the problem of leakage and accidental losses in refineries, which, when washed down the sewers, may result in additional contamination in the form of oil slicks. A generalized process flow diagram for a petroleum refinery is shown in Figure 4–9.

A coincidental problem with refinery operations is the occasional or accidental spillage of crude oil from tankers. These events often create toxic and extremely unsightly oil scums in the ocean and on beaches. Oil damage may reach staggering proportions and persist for long periods of time, for these wastes are extremely difficult to control.

Fruit and Vegetable Canning Wastes

Fruit and vegetable processing can be carried out in a single cannery, and by processing more than one food product the canning season for any one plant may be extended. The various processes involved in canning operations

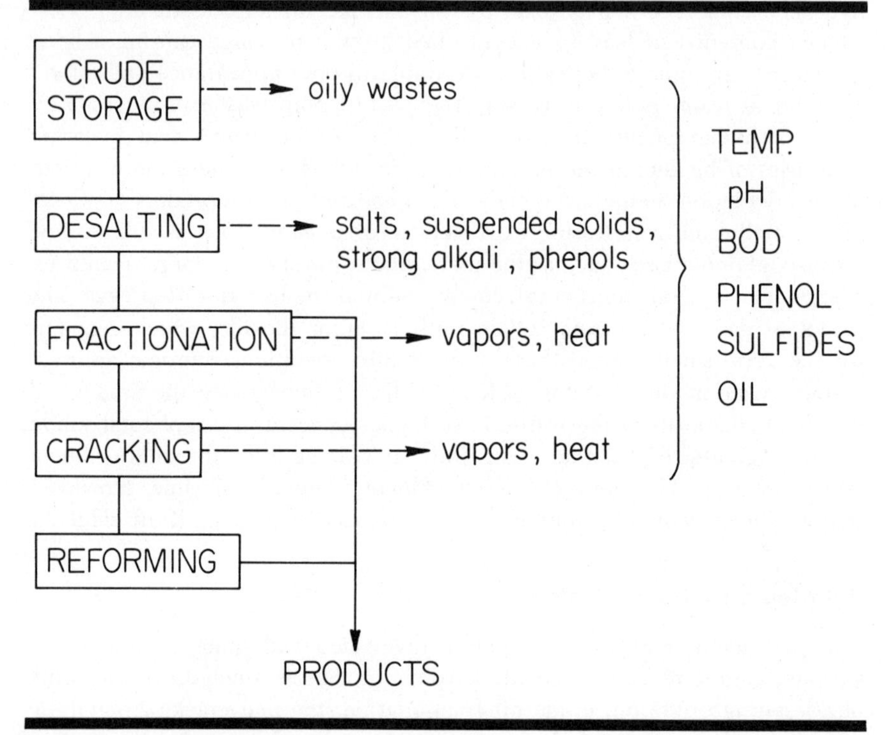

Figure 4–9. Petroleum Refining Process Flow Diagram

may be broken down into preparation, sorting and trimming, canning, cooling, and shipping. Wastes from food processing are not considered a public health hazard, but the liquid wastes are usually about 10 times as strong as domestic wastes in their demands on the oxygen supply in receiving waters. Also, the suspended solids of average canning wastes are relatively high compared to domestic sewage. Treatment of these wastes is required mainly to protect the oxygen level in receiving streams.

The disposal of wastes from food processing plants is an integral part of the total production system. See Figures 4–10 through 13. To reduce the costs of conventional waste disposal, recycling, reusing, and by-product recovery should be considered as alternatives to waste disposal. Efforts to reduce the total waste load often offset the cost of the necessary facility modifications. Since the capital and operational costs of waste treatment facilities are closely related to the volume of waste treated, reduced volumes can reduce the size of treatment units and the overall cost of treatment.

Studies of food processing plants reveal considerable variation in the strength and size of waste-streams. A considerable saving of water and waste can be achieved through good water management. In evaluating water reuse, the processor should consider counter-current flow. In this system, water

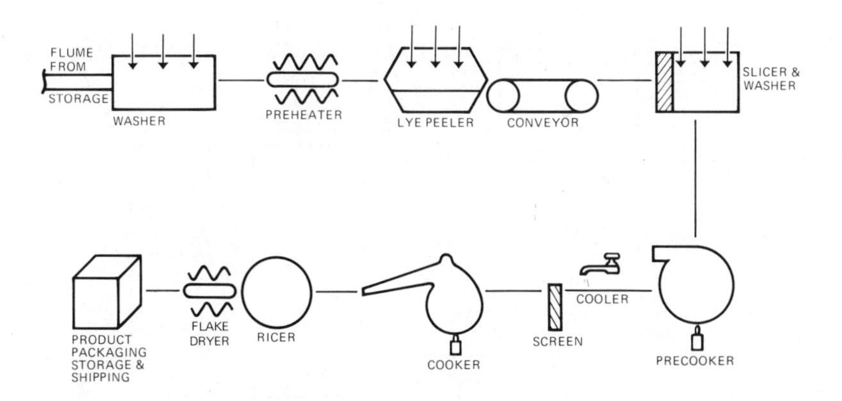

Figure 4-10. Typical Potato Flake Plant

CONTRIBUTING UNITS	EFFECTED CHANGE IN WATER QUALITY					
	Q	ΔCOD	ΔBOD	pH	ΔTot. Sol.	ΔVol. sol.
	gal.	lb.	lb.	initial 7.2	lb.	lb.
FLUME & WASHER 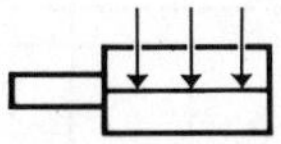	388	1.6	0.6	7.8	3.8	0.6
SPRAY PEELER & CONVEYOR 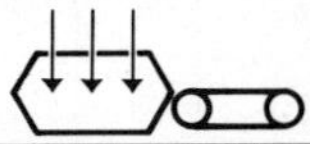	715	62.7	40.0	12.6	79.3	51.0
SLICER, WASHER, PRECOOKER & COOLER 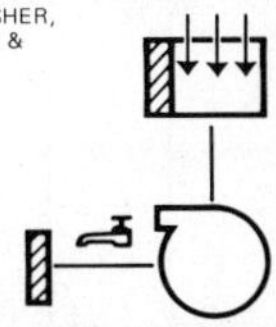	1540	50	38.4	5.2	45.1	42.1
TOTAL	2643	114.3	79.0	—	128.2	93.7

Figure 4-11. Major Liquid Waste Load Sources for Potato Flake Plants

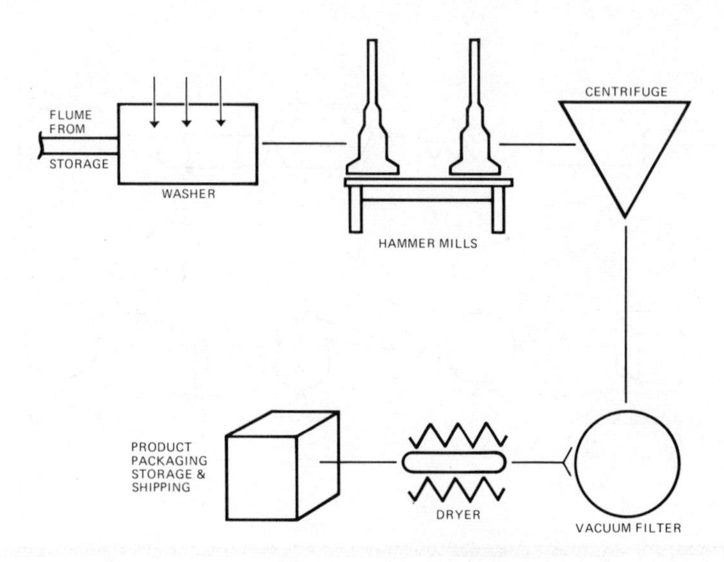

Figure 4–12. Typical Potato Starch Plant

CONTRIBUTING UNITS	EFFECTED CHANGE IN WATER QUALITY					
	Q gal.	ΔCOD lb.	ΔBOD lb.	pH	ΔTot. Sol. lb.	ΔVol. Sol. lb.
FLUME & WASHER 	560	46.2	17.5	6.8	60.2	41.2
HAMMER MILLS, CENTRIFUGE & FILTER 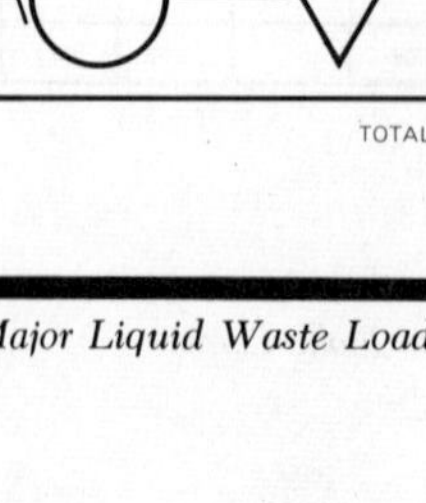	278	61.8	42.3	6.3	69.6	53.1
TOTAL	838	108	59.8		129.8	94.3

Figure 4–13. Major Liquid Waste Load Sources for Potato Starch Plants

from the last operation, instead of being wasted, is collected and passed back to be used in the preceding operation. The overall effect is that as the product is moved forward after each washing operation it moves into water that is cleaner than that used in the preceding operation.

Aluminum Refinery Wastes

Processing pure aluminum from bauxite ore is accomplished in two steps: first by removing the alumina ore from other mineral impurities, and second, by converting this mineral into metallic aluminum. The first step produces concentrated wastes which usually are confined to the mining regions. In this process sodium aluminate is shipped from bauxite mining regions in the southeastern United States and Australia to electrolytic smelters, many of which are located in the Pacific Northwest. Aluminum is then removed from a salt bath of alumina and extremely toxic fluorine salts which release fluorine gas to the air. The air above these electrolytic smelters is scrubbed, usually with water, to remove the toxic gas. Without further treatment, such a process can convert this air pollution problem into a significant water pollution problem. If water scrubbers are employed to treat the gases above the electrolytic smelters, the waste stream will include significant concentrations of fluoride ion and particulate alumina particles in a generally acid solution. In addition, there may be quantities of carbon in the waste waters, since these plants often have anode curing ovens which produce electrolytic anodes used in the smelting operations.

SUMMARY

Waste is an unavoidable consequence of satisfying man's needs for food, water, air, space, shelter, and mobility. In any material process, by-product recovery or recycling can substantially alter waste quantity and quality, but all processes eventually produce some waste.

Energy is a prerequisite of all human activity, and often is required at locations chosen by man in greater intensity and for longer durations than is available naturally. To meet his energy requirements, man has developed the technology to use energy stored and concentrated in the limited resource forms of water power, chemicals, and fossil and nuclear fuels, and through such mechanical devices as the combustion chamber, the engine, and the turbine. Waste by-products of such intensive energy conversion practices include ashes, gaseous emissions, heat, radioactive emissions, and noise. The environmental impacts of waste often are magnified by virtue of increased human densities resulting from urbanization, the net effect of which is not only increased domestic waste, but decreased areas of the natural environment available for waste discharge. Agriculture is responsible for most of the annual production of solid wastes in the United States, and irrigation is the single largest consumptive water use. Industrial activity, like agricultural processes, constitutes a major use of the water resources of the United States, but fortunately, it is not a major consumptive use. It is not possible to generalize in depth regarding the composition of industrial wastes because of the tremendous variety of activities subsumed under the name "industry." Various industrial processes produce a variety of materials in their industrial waste streams and a holistic perspective must be adopted in attempts to control such wastes; i.e., solid materials removed in treatment of waste water streams can in themselves become objectionable solid wastes, while gaseous components in exhaust removed by water scrubbing may create potentially serious water pollution problems. Because wastes are inevitable products of industry, judicious selection of processes and products must be made to optimize the use of our resources and to provide the least possible insult to the environment.

References—Chapter 4

1. Wolman, A.
"The Metabolism of Cities," *Scientific American*, September 1965.

2. Davis, Kingsley.
"The Urbanization of the Human Population," *Scientific American*, September 1965.

5 Effects of Wastes on Aquatic Communities

ENRICHMENT OF ECOSYSTEMS

Since man is a member of many ecosystems and is very effective in manipulating the resources of each system, his activities can alter the quality, quantity, and distribution of materials and energy in the system. Ecosystems depend upon the uninterrupted cycling of biologically important materials and require the continuous input of energy. By adding wastes that contain energy and nutrients, ecosystems can become locally enriched and their productivity may increase significantly.

Ecosystems can be enriched by organic or inorganic wastes. Organic matter is found in wastes from domestic sewage, pulp mills, meat packing plants, fish canneries, sugar beet plants, barnyards, and other sources. Organic components contain energy that is utilized by heterotrophic microorganisms. Energy can be transferred from these organisms to other members of the system. Instead of enlarging the energy source for heterotrophic microorganisms, inorganic nutrients stimulate the growth of autotrophic green plants, which get their energy from sunlight. Inorganic nutrients such as nitrogen and phosphorus can come from domestic sewage, agricultural fertilizers, detergents, soil erosion, rainfall, and runoff. Although some nutrient enrichment problems in urban areas result from domestic sewage, nutrient enrichment problems have diverse sources. Of the 5 billion pounds of nitrogen and 1 billion pounds of phosphorus that enter the nation's waterways annually, from 2 to 17 billion pounds of nitrogen and 0.3 to 2 billion pounds of phosphorus can be traced to rural runoff. This is two to ten times the nitrogen and one and a half to four times the phosphorus from domestic sewage.[1]

Regardless of whether enrichment comes from organic matter or inorganic nutrients, the effects on ecosystems are similar. Either pathway—i.e., sunlight and inorganic nutrients to primary producers (autotrophic green plants), or organic matter to decomposers (heterotrophic bacteria and fungi)—can increase the productivity and biomass of fish and higher aquatic life. See Figure 5–1. Moderate, controlled enrichment can be beneficial to man, since it can lead to increased production of desirable species. It can actually help increase food supplies that would otherwise remain relatively stable. If enrichment is excessive, degraded water quality can result. Dissolved oxygen can be depleted so that only nuisance-causing and otherwise unwanted forms of life remain. The question of how much enrichment is too much is a difficult one to answer. A more detailed discussion of the effects of enrichment and

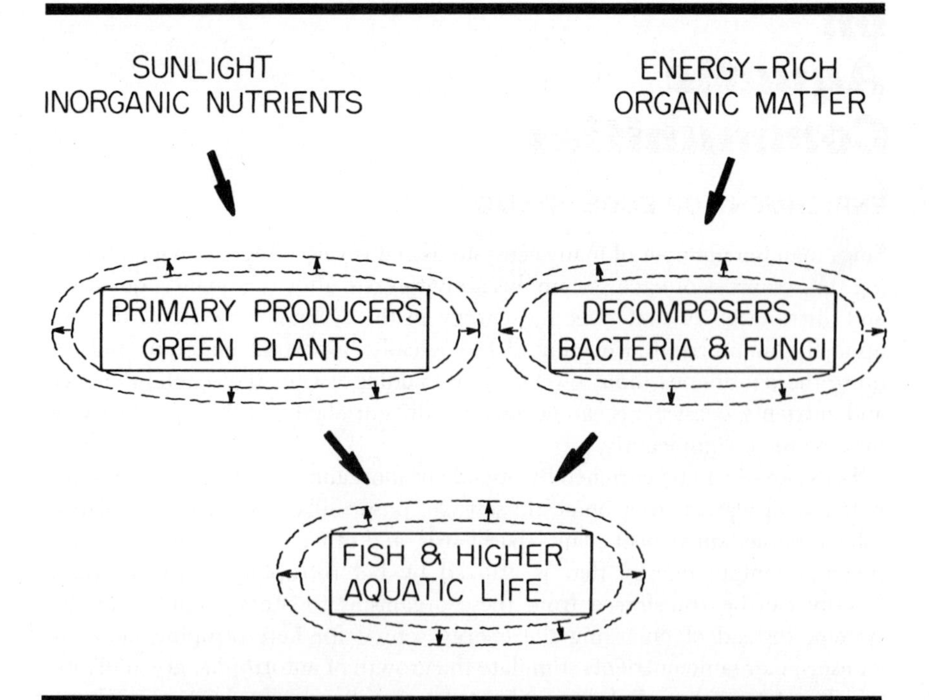

Figure 5–1. Alternate Pathways of Enrichment

The potential increase in production at simplified levels in the food chain as a result of enrichment by either inorganic nutrients or energy-rich organic matter.

the present level of understanding of the process should enable the reader to appreciate why a very conservative approach to enrichment control is necessary.

Eutrophication

Bodies of water age naturally. Over thousands of years, nutrients are contributed to lakes directly by rainfall, ground and surface water flow, and dissolution from the lake bed. Even when the rate of natural nutrient addition remains relatively constant, lakes tend to gather organic material. As a result, the effective volume of the lake decreases and the concentration of nutrients increases, a process known as *eutrophication.* Eutrophication also occurs in running water insofar as nutrient enrichment is concerned, but streams do not age as lakes do. Eutrophication of aquatic ecosystems occurs naturally, therefore, although man can greatly accelerate the process.

As concentrations of nutrients (particularly nitrogen and phosphorus) increase, the productivity of autotrophic green plants also increases. In most natural waters, the limited availability of either or both of these nutrients restricts the growth of green plants. If enrichment reaches high levels, plant growth may no longer be limited by nutrients. As the organisms become more dense, they block out sunlight and the process is reversed.

The consequences of increased growth in enriched waters are far reaching. The biomass of dead and dying plants and animals settles to the bottom, along with organic and inorganic materials that wash in from the watershed, and helps fill the lake with sediment. As the lake becomes shallower, the water is more completely mixed by the wind and heated more evenly. Better mixing further enhances eutrophication by increasing the temperature and by making nutrients in the sediments more accessible.

Nutrient enrichment can also have an impact on the water's oxygen resources. As the organisms produced in surface waters die, they settle to the bottom and decay. Decomposition requires oxygen, the demand for which increases as the concentration of available nutrients increases. When lakes stratify thermally during summer, vertical mixing of bottom water is retarded. If the oxygen demand created by decaying organic matter exceeds the oxygen supply, depletion may occur. The oxygen supply may be entirely depleted if stratification is strong and the oxygen demand exceeds the initial supply in the deep waters of lakes. Reduced oxygen levels in deep, cool waters can force cold-water fish, such as trout and whitefish, to surface waters for oxygen and can cause their elimination in favor of warm-water fish. In much the same way, other species that are intolerant of changing environmental conditions can be inhibited while more tolerant species can be favored. Many of these tolerant species create nuisances either through sheer abundance or because they have unusual growth characteristics. One of the principal consequences of eutrophication is dominance by blue-green algae. Massive blooms of blue-green algae are nuisances because they float and accumulate at the surface. Such mats are ugly, form slimy coatings on bathers, and create undesirable taste and odors in drinking water.

Large, rooted aquatic plants create nuisances for bathers, fishermen, and water supply intake facilities, and generally degrade water in appearance. Much of the animal and plant material produced at the surface will sink to the bottom, increasing the organic content of the sediment and creating an ideal substrate for the optimum growth of rooted plants in shallow areas of lakes. Rooted aquatic plants will continue to propagate, and as the lake becomes even shallower, the zone of plant growth will move toward the center until the lake is entirely covered by plants. The plants themselves act as sediment traps, so the lake will eventually fill completely and become a terrestrial environment.

In summary, the natural process of eutrophication can be described by the change in general characteristics as shown in Figure 5–2. Productivity increases gradually through the mesotrophic stage, then increases more rapidly as eutrophy advances. The variety of life forms decreases rapidly in the eutrophic stage, although dissolved materials increase at about the same rate throughout the aging process. Undesirable fish species tend to replace desirable species in the eutrophic stage. If lakes could be maintained in the mesotrophic stage, productivity would remain moderately high, the variety of life forms would not be reduced, and desirable species would predominate.

The rate of aging of any lake varies according to the quantity of nutrients it receives, the rate at which its water is exchanged, its depth, and other factors. Increased nutrient input will not stimulate the aging process in a deep lake as much as it will in a shallow lake. Aging is also slower in a lake with a rapid water exchange rate. A shallow lake with slow water exchange is very sensitive to increased nutrient input. It is important to emphasize that the precise relationship of these three factors as they relate to the rate of aging is not well studied. Thus, to control eutrophication and hold lakes in a desired stage such as mesotrophy will require a much more extensive scientific understanding of the process than presently exists. Until this knowledge is acquired, artificial enrichment of aquatic ecosystems, or "cultural eutrophication," must be minimized as much as possible. There is a real danger that, due to our incomplete understanding, we may overshoot the desired productivity mark and find it impossible to reverse the process.

Our incomplete understanding of eutrophication is illustrated by the carbon-phosphorus controversy that has arisen in the past couple of years. The nutrient present in the smallest concentration relative to the metabolic requirements of, for example, algae, is referred to as the limiting nutrient. One nutrient usually does not limit growth in all bodies of water and certainly does not limit growth all the time in any one body of water. Because the phosphorus supply is most limited compared to the other two major nutrients (nitrogen and carbon), it is usually considered the limiting nutrient. Nitrogen and carbon can enter aquatic ecosystems from the atmosphere, but phosphorus must originate from geological formations. However, to build organism biomass, carbon and nitrogen are required in larger amounts than phosphorus. Based on cell content of carbon, Kuentzel[2] has calculated that for large algal blooms to occur, outside sources of carbon (e.g., from sewage) are needed. Carbon has also been shown to limit algal growth in soft waters with low carbon content.[3]

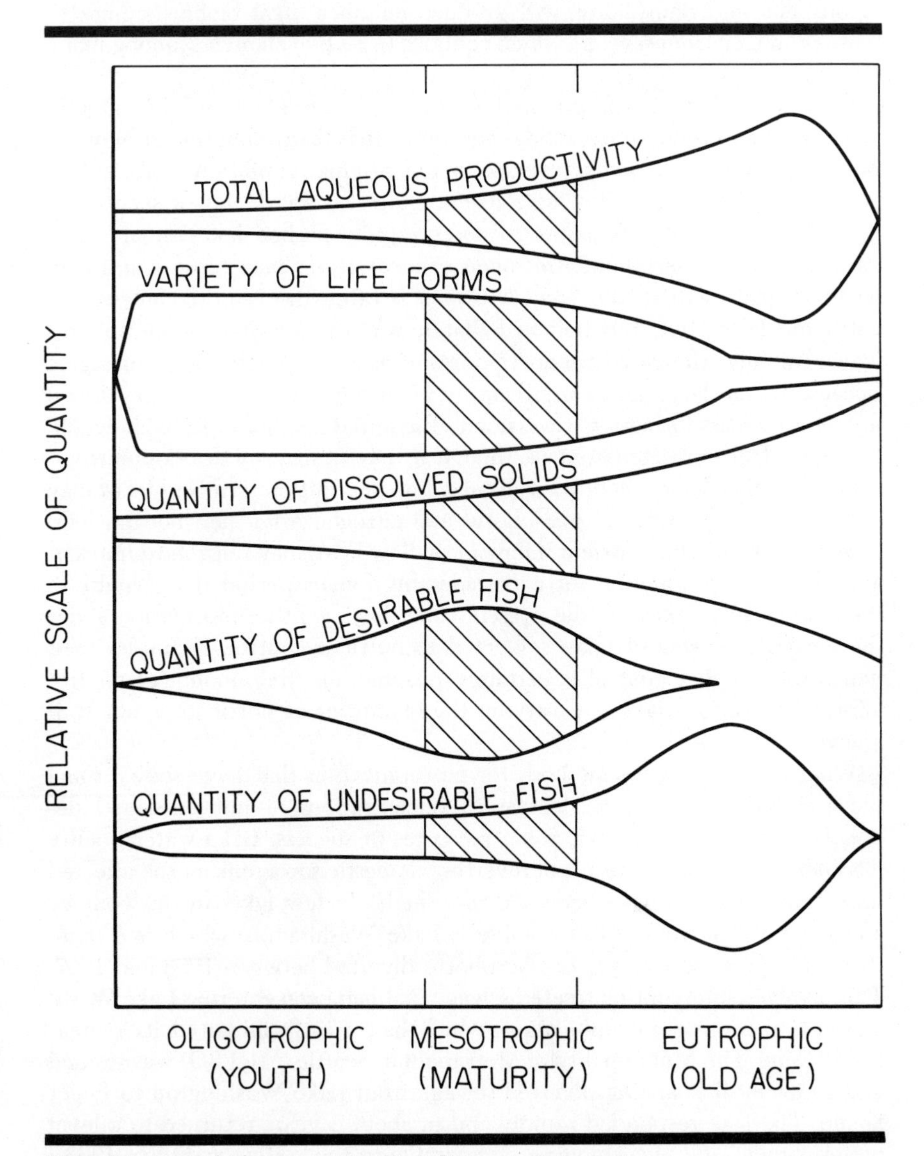

Figure 5–2. Typical Changes in Atrophic Indicators in Aging Lakes

From LAKE ERIE REPORT—A PLAN FOR WATER POLLUTION CONTROL, U.S. Department of Interior, FWPCA, Great Lakes Region, 1968. Reprinted by permission of the U.S. Environmental Protection Agency, Region V (successor agency to FWPCA GLR via reorganization plan No. 3 of 1970).

Most scientific information, however, supports the contention that usually phosphorus, but sometimes nitrogen, is the limiting nutrient in lakes. Simple laboratory experiments show that when nitrogen and phosphorus are added to flasks containing lake water, large masses of algae bloom in a short time. Small increases (if any) occur in control flasks without these nutrients added. Further, work in the artificial fertilization of lakes has shown that the addition of nitrogen and phosphorus will produce nuisance algal blooms similar to those produced from sewage, which contains these two elements among many others.

Because adding phosphorus and nitrogen to unproductive lakes greatly increases their productivity, it does not follow that their reduction or elimination will result in a quick reversal to the original condition. Information on such reversal rates following nutrient control is scarce, but it seems that reduction of phosphorus inflow to deep, rapidly flushed lakes should have the most merit, even though nitrogen concentrations may be low and may be limiting growth in the lake. Nitrogen is often the limiting nutrient in eutrophic lakes, but it is believed that it was the addition of phosphorus which initially stimulated production, resulting in the depletion of nitrogen. Because of the large and varied supply of nitrogen which could be drawn upon for biomass formation, reversion to the initial condition probably would not result from nitrogen control, though it may limit growth temporarily.

On the other hand, shallow, well mixed or intermittently mixed lakes may receive a sizeable input of phosphorus and nitrogen from their bottom sediments.[4] Reducing the nutrient inflow in shallow lakes may not retard nutrient availability or production for a considerably longer period than would be the case in deep lakes. At the present time there is little information about the dynamic aspects of lakes subjected to nutrient control. Examples exist which suggest that control is certainly possible, but the amount of control afforded in various lakes by varying the quantities of nutrients is not well known.

What are the control methods for eutrophication that have shown some success? Increased clarity, decreased concentrations of nutrients, and decreased biomass of algae are some measures of success. Lake water quality has been improved as a result of diverting domestic sewage from the affected lake. Improvements have been accomplished in a few lakes in this country and several in Europe. Most notable is Lake Washington, which had more than half of its total supply of phosphorus diverted between 1963 and 1967. This involved intercepting treated sewage that had been entering Lake Washington and causing eutrophication, which had greatly impaired its recreational value. The Municipality of Metropolitan Seattle (METRO) was formed and under local financing diverted sewage from Lake Washington to Puget Sound. The lake responded rapidly and in about 3 years returned to a level of enrichment and algal biomass comparable to the water quality of 1933.[5]

One might wonder what will happen to Puget Sound—will it become just another but larger Lake Washington? As is typical with ecological problems, this question cannot be answered at this time. Work by Dr. George Anderson at the University of Washington shows that algal production in Puget Sound is limited by available light, but when vertical mixing of the water is least more light is available to the algae and production is increased. Another important point is that when Lake Washington was receiving its greatest enrichment, the maximum phosphorus concentration was about the same as Puget Sound contains naturally. From all indications there appears to be no immediate threat of creating a problem in Puget Sound by the addition of sewage. However, continual and complete monitoring of the waters of Puget Sound is necessary in order to detect early signs of change in time to take action and avert undesirable consequences.

The case of Lake Washington illustrates clearly that deep lakes (mean depth 37 meters) whose water is exchanged quite rapidly (about 3 years) by low-nutrient inflow water, can rapidly recover from eutrophication. The Lake Washington case further illustrates that phosphorus was the important ingredient in sewage that caused the lake's premature aging. However, there are many unanswered questions about the effects of phosphorus and sewage diversion in shallow lakes and the specific causes of the nuisance algal formation. In a case similar to Lake Washington, about half of the incoming phosphorus was intercepted and diverted from Lake Sammamish in 1968 by METRO to prevent that lake from reaching a trophic stage as advanced as that of Lake Washington prior to diversion. Lake Sammamish is smaller and shallower than Lake Washington: a quarter of the area and half the depth. Three years later this lake had not shown a significant reduction in its nutrient content or algal biomass, or an increase in clarity.[6] The lack of evidence of change poses significant questions.

Another control method for eutrophication is tertiary sewage treatment—removal of most of the phosphorus and/or nitrogen from domestic waste water. This technique and its costs will be discussed in the chapter on control, but here we can indicate the potential success of such a method. Laboratory experiments indicate that tertiary treatment negates most of the stimulatory effect of treated sewage on some species of algae.[7] Sewage effluent that once entered Lake Tahoe now receives tertiary treatment before it is discharged into Indian Creek Reservoir and has not created an undesirable effect in that reservoir.[8] Although actual test cases of the potential for tertiary treatment in eutrophication control are scarce, it seems that nutrient removal where conditions warrant is the best approach to solving this important and growing problem. A sobering thought, however, is that although tertiary treatment removes 90 percent of phosphorus, the amount discharged remains constant as human populations and their waste loading increase. Thus, the actual quantity of phosphorus discharged into the nation's waters will continue to increase. The fact that phosphorus is trapped in lakes and has the potential for continuous recycling in shallow lakes suggests that in some situations tertiary treatment may not be a permanent solution and may need to be augmented with some other measure.

Proposals to control eutrophication in Lake Erie include phosphorus control by tertiary treatment, and elimination of phosphates from detergents. Substantial reduction of phosphorus inflow could be accomplished by removing 80 percent of the phosphorus through tertiary treatment of domestic waste waters and the elimination of phosphates in detergents. Because of the shallowness of Lake Erie and potential phosphorus recycling from sediments, there is some doubt that even tertiary control would rapidly reverse eutrophication.

Other controls for eutrophication that have also shown some success are adding low-nutrient dilution water, dredging enriched sediments, and adding chemical complexing agents such as aluminum ions and fly ash. Adding low-nutrient dilution water reduces the concentration of limiting nutrient, thus reducing the algal growth rate and ultimate biomass. If the water exchange rate is rapid enough, algal biomass can be reduced by washing algae from the system faster than it can be replaced through growth. Dilution has shown some success in Green Lake, a naturally eutrophic lake in Seattle,[9] and holds some promise for Moses Lake, a hypereutrophic lake in eastern Washington.[10]

Organic Enrichment

When organic matter (sewage, pulp mill waste, meat processing waste, etc.) enters a lake or stream it is said to exert an oxygen demand. This oxygen demand can be measured and is referred to as the biochemical oxygen demand (BOD). Depending upon the amount of dilution water available, the rate of re-aeration of the water from the atmosphere, and aquatic plant photosynthesis, the water's oxygen content may be reduced only slightly or it may be completely used up. The change in oxygen content of a small stream receiving sewage was one of the first effects of pollution carefully documented and was hypothetically illustrated by Bartsch and Ingram, as shown in Figure 5–3. Changes in the biological community of organisms accompany changes in oxygen downstream from sources of organic matter. Changes in the community occur because certain organisms need fairly high concentrations of oxygen; trout and other game fishes, and most of the aquatic insects that serve as fish food need high oxygen concentrations. Very low levels of oxygen can be tolerated by others, such as rough fish and a few of the very tolerant insects, worms, and air-breathing snails. Therefore, when oxygen-consuming waste is added to a stream and the oxygen content drops, only the tolerant organisms survive. Since there are relatively few tolerant species, they generally reach fairly large population densities. See Figure 5–4. The great abundance of tolerant organisms in highly polluted streams can serve as significant food supplies for those fish that do survive. In some instances fish move into these areas of rich food supply during the day when photosynthesis has added enough oxygen to render the waters tolerable. At least 5 milligrams of dissolved oxygen per liter is generally recognized as a minimum for the survival and growth of most fish species. In waters contain-

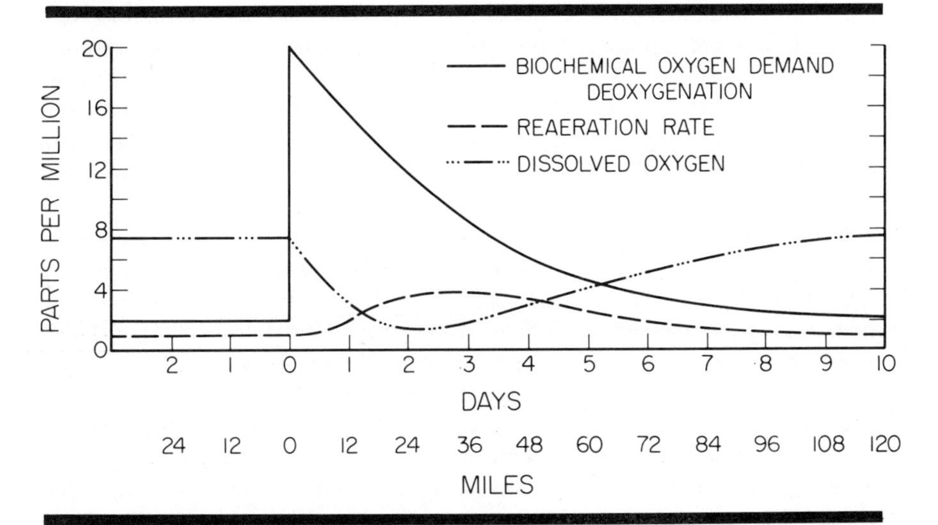

Figure 5–3. Effect of Domestic Waste on Downstream Water Quality (assumed conditions—river 100 cfs and population 40,000)

From STREAM LIFE AND THE POLLUTION ENVIRONMENT, by A. F. Bartsch and W. Ingram. PUBLIC WORKS, Vol. 90, 1959. Reprinted by permission of the publisher.

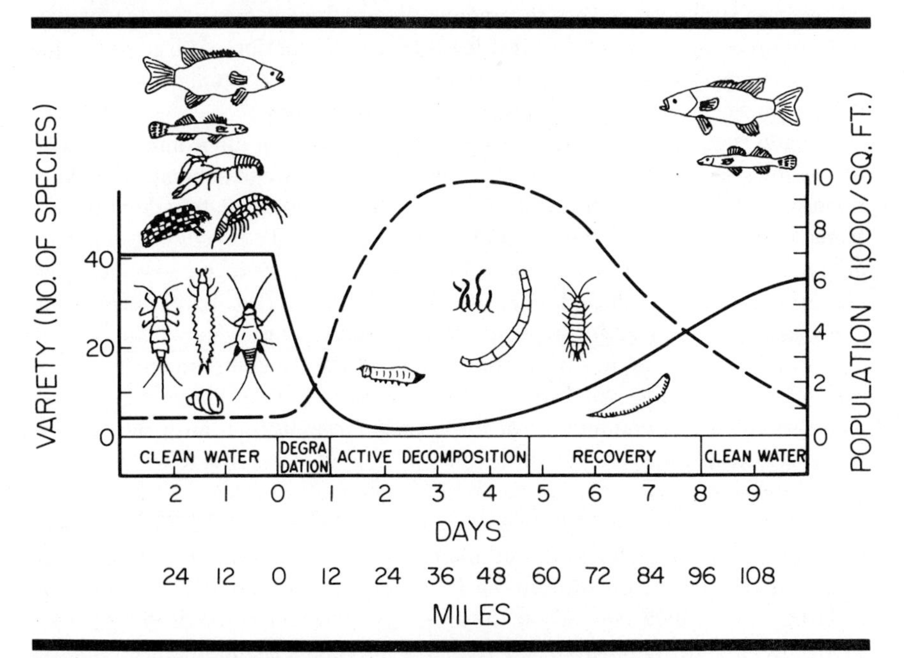

Figure 5–4. Effect of Domestic Waste on Downstream Water Quality

The change in biological communities with distance, as indicated by species number and abundance in a hypothetical stream (100 cfs) receiving an input of domestic waste from a community of 40,000.

From STREAM LIFE AND THE POLLUTION ENVIRONMENT, by A. F. Bartsch and W. Ingram. PUBLIC WORKS, Vol. 90, 1959. Reprinted by permission of the publisher.

ing trout, 7 mg per liter is necessary. The difficulty in setting standards is that fish do not require any one level of oxygen all the time. For example, a fish species can get along quite well on 2 to 3 mg of oxygen per liter when it is not growing, reproducing, or working too hard, but at times when it is very active it may need as much as 8 or 9 milligrams per liter. The only concentration that is safe at all times is atmospheric saturation, 9 mg per liter at 68°F.[11]

Organic waste may not be the only source of oxygen-consuming material resulting in fish mortalities or community shifts to undesirable organisms. Natural conditions can develop in water, such as the production of large masses of algae due to the inflow of organic material from land surfaces, which can also reduce the oxygen content. In many natural lakes a fine line may exist between rich, productive water yielding large game fish and waters that will not produce game fish at all because of too much richness resulting in persistent low oxygen concentrations. A condition referred to as winterkill can occur in rich lakes during the winter. Under such conditions there is no re-aeration of oxygen from the atmosphere, and with a large amount of organic material the demand for oxygen can be greater than the oxygen supply. Thus, part way through the winter a fish kill may occur, and even though such a lake is normally very productive, it can quite often be fishless. However, a certain amount of enrichment from sewage or any other source can be beneficial, if it leads to the production of large, desirable species of game fish but does not reduce the oxygen content to a seriously low level. Such water could hardly be called polluted.

Oxygen reduction is not the only effect on the stream environment caused by organic wastes. Particulate matter suspended in sewage effluents can affect the distribution of algae. See Figure 5–5. In the area immediately downstream from the outfall, where the particulate matter is highly concentrated before a significant amount has settled, light penetration is reduced so much that algae cannot grow. Once the particulate matter settles, light penetration can again be sufficient for algal growth. In the examples shown in Figures 5–3 and 5–5 this may not occur until most of the oxygen demand has been satisfied some 70 miles downstream from the waste source. The settled matter can also affect the community of organisms. Orgamisms preferring rich organic sediments, such as worms, can develop in great numbers, although organisms such as aquatic insects, which prefer clean stream beds and like to live in the open spaces between rocks, can be eliminated because these open spaces tend to fill with sediment, as shown in Figure 5–4. Organic sediment or sludge can literally seal the stream bottom; the reproduction of fish species which depend on the free flow of water through gravel, such as trout, may be eliminated. In the area of high organic-matter content, bacterial slime growths can develop, smothering incubating fish eggs and many aquatic insects. However, it has been shown that even obnoxious-appearing bacterial slime growth can provide a food supply for those insects

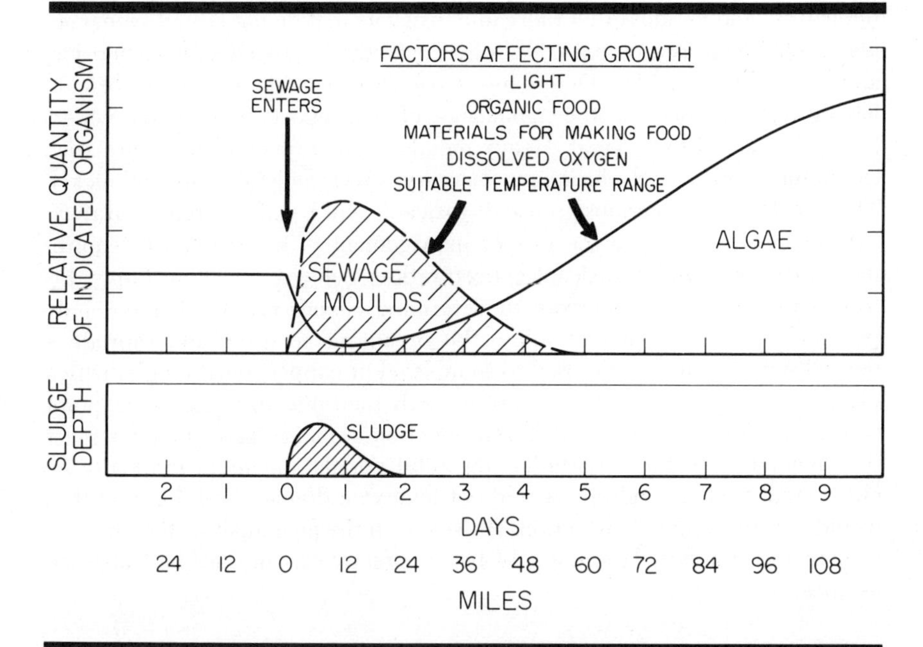

Figure 5-5. Effect of Domestic Waste on Downstream Water Quality

The change in algae, sludge accumulation, and moulds with distance in a hypothetical stream (100 cfs) receiving an input of domestic waste from a community of 40,000. The sludge is decomposed gradually; as the water clears algae begin to grow.

From STREAM LIFE AND THE POLLUTION ENVIRONMENT, by A. F. Bartsch and W. Ingram. PUBLIC WORKS, Vol. 90, 1959. Reprinted by permission of the publisher.

that are not eliminated. If oxygen reduction and sedimentation have not advanced to the point where these factors interfere greatly with fish survival, then fish species as sensitive as trout can take advantage of an increased food supply from the bacterial slime growths and show extremely high growth rates.[12] Here again, a little organic enrichment may not necessarily affect ecosystem function adversely.

INHIBITION EFFECTS

In contrast to wastes that have enriching potential, many wastes either are directly toxic or in other ways inhibit life systems. Toxicity is a chemical property that adversely affects various life processes so that death may result. Many chemicals have specific modes of toxicity; some are irritants that affect all cells alike and others act internally on respiratory or other functions. Almost any chemical is toxic in some amount; this amount is reached in an organism when the intake rate exceeds the excretion rate. Groups of chemicals that are toxic to fish and aquatic life in general include the heavy metals, organo-pesticides (slimicides, rodenticides, herbicides, fungicides, and fumigants), and various other inorganic toxicants that do not fall into a particular group such as ammonia, hydrogen cyanide, hydrogen sulfide, fluoride, and arsenical pesticides. Detergents have also been shown to be toxic in laboratory situations, but the significance of their effects in receiving waters is not considered to be great. Heavy metals of greatest significance are zinc, cadmium, copper, lead, chromium, mercury, silver, molybdenum, and nickel. The effects of these various groups of toxicants will be considered separately.

Control of wastes and the proper management of water quality requires more than maintaining toxicant concentrations below their lethal limits. Efforts are continually underway to determine concentrations that not only prevent mortalities but also allow the organisms to grow and reproduce normally. Fish could be exposed to some level of copper, for example, which did not cause death but did interfere with their life processes to a point where growth was inhibited; if reproduction became inadequate to meet the demands of natural mortality, the total population might gradually be eliminated. Such a condition would not involve a dramatic fish kill and thus would not attract much attention. However, in the final analysis the elimination of the fish population would be as effective as if they had all died at once.

Acid Mine Wastes

Mining activities are an important source of some metallic wastes. A brief discussion of acid mine drainage will illustrate the interaction of various chemical factors on the toxic potential of a metal. This problem develops because the mined-out portions of coal veins fill with water and, in the absence of oxygen, reduced sulfur compounds are formed. Biologically mediated oxidation of these sulfur compounds forms sulfuric acid and consequently produces a very low pH in the waste water. This is important because very high concentrations of metals such as copper, zinc, and iron are held in aqueous solution under low pH conditions. The addition of these waste waters to receiving streams results in extremely toxic conditions and a lifeless environment. Serious conditions can often develop when a mining company experiences a labor strike. During strikes, acid waters accumulate, the waste discharges are not treated to reduce acidity, and when work finally commences the accumulations of acid water are flushed out of the mines and may create extensive fish kills.

Pesticides

Some of the most toxic chemicals known to man are the organo-phosphorus and chlorinated-hydrocarbon insecticides. Endrin, a chlorinated hydrocarbon, is lethal to many species of fish at concentrations as low as fractions of a part per billion. Some of the organo-phosphorus compounds are nearly as toxic, although they break down in water and do not cause the serious residual effects that are attributable to chlorinated hydrocarbons. One cannot stress the significance of these pesticides too strongly in view of their increased use and the resulting contamination of the environment. Although a considerable amount of research has been done on the effects of some of these pesticides, new pesticides are developed faster than their effects can be determined. In no case is the total effect of any pesticide known well enough to determine an accurate, safe level for a water system. A major difficulty here is that toxic effects may be cumulative and the residues build up to dangerous concentrations in the body. The cumulative effects of pesticides on reproduction in birds and other wildlife have been known for some time, but it has been only recently that detrimental effects on reproduction in aquatic organisms have been demonstrated.[13] More research is needed to delineate the precise effects of cumulative pesticide residues on the physiological functions of aquatic organisms. However, even if safe levels were known, the sources of pesticides are so diverse and they already are so well distributed throughout the environment that control short of total elimination seems impossible.

The persistence of chlorinated hydrocarbon insecticides is a particularly important environmental threat. Some of them, for example, have a half-life in soil of 15 years; that is, in 15 years half of the original material has changed to some other material that is no longer toxic. DDT, benzene hexachloride, aldrin, dieldrin, chlordane, and endrin are all chlorinated hydrocarbons. These compounds are stored in fatty tissues, but they can also be stored in fairly high concentrations in the brain or gonads, and to a lesser extent in other organs. Some are known to accumulate in carnivorous organisms at the top of food chains, in levels up to thousands of parts per million.[14] They are also absorbed through the body surface of animals because of the ease with which they pass through cell walls. While these substances are quite soluble in the fatty tissues of animals, they are fortunately almost insoluble in water. It is, therefore, unusual to find these pesticides dissolved in water at concentrations much greater than a few parts per billion or even per trillion. If organisms at the top of the food chain such as birds and some fish are able to concentrate pesticides to levels of thousands of parts per million when the concentrations in water are only a few parts per billion, then the factor of concentration is something in excess of 100,000 to 1.[15]

What consequences will this have on humans consuming the contaminated organism for food or on the survival of the organism itself and its population? For example, eating fish with high concentrations of insecticide in their tissues would mean that the total amount of material is liberated at one time and could potentially have a very significant effect on an individual. The threat is great since some fish have developed resistances to certain pesticides and have developed strains capable of absorbing high levels of insecticides in their tissues. It is likely that as fish are gradually exposed to greater concentrations of insecticides, those that can resist the higher concentrations are selected and will eventually dominate the population. The difference is that man is not gradually exposed to higher concentrations, as the fish are, but receives a large dose when eating the fish. Because of this threat, even though direct proof of damage to humans is generally lacking, the Food and Drug Administration (FDA) has set tolerance limits for pesticides in foods in the United States. If food contains a particular insecticide above the tolerance determined for that insecticide it cannot be marketed. The Food and Drug Administration ruling has affected recreational fishing recently in Michigan (1969), in a case dealing with the coho salmon. The DDT content in a sample of these fish was found to be about 19 parts per million (ppm), considerably higher than the FDA tolerance limit of 5 ppm, and the fish were confiscated. The tolerance limit for dieldrin set by the FDA is 0.3 ppm. DDT has been found in fish from all of the 50 nationwide monitoring stations in concentrations ranging from 0.03 to 57.8 ppm.[16]

The buildup of insecticide residues in food webs can also have a serious effect on the organisms themselves. It is useful to think of the known effects on the biota as a safety factor for the human population. The reproductive ability of the coho salmon in Lake Michigan has been shown to be much below normal, as indicated by the death of approximately half of the young in hatcheries.[17]

Insecticide that is accumulated by the adult female is passed on to the eggs and stored in the fat. When the young fish begin to develop they draw on the fat, and abnormalities develop in the hatching larvae. Survival rates of these abnormal fry are poor. The decline of fish populations in other bodies of water has been attributed to this same effect. Instances are known in which mortalities of fish following spraying for forest insects were delayed until the time of spawning, when the residues stored in body fat were probably released because of the stresses imposed on the fish at that time. Instances of mortalities that have resulted from pesticides are abundant, but examples of long-term effects from accumulations within the food chain are scarce. Painstaking observations are necessary to document long-term effects of sublethal concentrations in these situations, since it is necessary to achieve sampling continuity over long periods, and many organisms are difficult to observe and control in the environment.

Concentrations of dieldrin and DDT in fish in Lake Michigan are presently at the danger level.[18] Danger levels have not been reached in Lake Superior and Lake Erie, where fish contain from two to seven times less DDT than in Lake Michigan. The lack of surrounding agricultural land from which pesticides can drain is probably the reason why pesticides have not reached high levels in Lake Superior. However, why they do not appear as a significant problem in Lake Erie, which is much more polluted than either Michigan or Superior, is not clear.

All of us who use pesticides, regardless of the frequency and amount, contribute in measurable ways to the overall problem. Pesticides can be distributed quite easily through the air, via mobile animals in the food chain, and carried with either water flowing in channels or runoff from land. Once dispersed in the environment, they remain for long periods of time, and it is for this reason that pesticides are used widely to control insects. Pesticide users want something that is inexpensive and lasting. It is likely that the hard pesticides will be used for some time to come. There seems to be no method for controlling pesticides in the environment, because of the capacity of organisms to absorb them even from very small ambient concentrations. Therefore, the only alternative seems to be to stop using hard pesticides until more pest-specific compounds are found. A nationwide ban on the use of DDT is a step in that direction. This, however, might greatly reduce agricultural production for a time.

Radionuclides

In contrast to pesticides, which are accumulated without any apparent metabolic use, many radioactive elements are accumulated either because the element has some metabolic use, or because it simulates the use of a like element. Cesium 137, strontium 90, and iodine 131, which originate largely from atomic fallout, simulate the activity of potassium and calcium, respectively, in the human body. Cesium 137 is distributed throughout the organisms, strontium 90 concentrates in bone tissue, and iodine 131 ends up in the thyroid gland and is concentrated to relatively high levels. Phosphorus is an important constituent of living cells, and phosphorus 32 can be concentrated in humans from a hundred to many thousands of times its concentration in water.

Other food chains besides the aquatic chain have man as the last link. Insecticides and radioactivity may reach man either by direct application, or by fallout onto agricultural crops and vegetation that is consumed by domestic livestock. Many of the sources for these contaminants are monitored, but determining tolerance levels for man is a difficult proposition. Knowing the concentration in the food source is only part of the problem. The final concentration in a consumer organism, such as man, depends on the organism's consumption and excretion rates, in addition to the metabolism of the contaminant within the body. It is difficult to set tolerance limits on poisons that have such a dynamic character in biosystems. The problem is compounded because of our understandable reluctance to experimentally evaluate the effects of these insidious materials on human beings. Most of the scientific data which do exist were obtained as a result of accident or tragedy.

Heavy Metals

Other substances that can accumulate to high concentrations in the food chain are heavy metals such as mercury and lead. A classic example of mercury poisoning through food chain buildup occurred in Japan's Minimata Bay, which received industrial wastes containing mercury compounds. Fish in the bay accumulated mercury compounds through the food chain and were in turn caught and eaten by people, as well as by alley cats that frequented the dock area. During 1953–1961, 44 people died from mercury poisoning as a result of eating fish caught in the bay.[19] Today, local residents have learned to watch the cat population. Cats are apparently more susceptible than humans, so when the cats on the docks begin to die people stop eating fish caught in the bay.

Mercury residues, particularly in commercial food species, have recently been recognized as a worldwide problem. During 1970, 20 states in the United States banned commercial fishing because mercury concentrations exceeded the 0.5 ppm FDA limit. Control since the discovery of this widespread problem has significantly reduced mercury discharges. The problem, however, may require more than curtailment of discharges. The principal deleterious effect of mercury is caused by its organic derivatives—dimethyl and monomethyl mercury. These compounds are produced by microorganisms in the sediments of lakes and rivers via the reduction of elemental mercury. Elemental mercury is not soluble in water, nor is it easily absorbed by organisms. The organic forms, particularly monomethyl mercury, are readily absorbed by organisms and can thus be concentrated to high levels. Sediments in many areas are highly saturated with elemental mercury, and may require many years to be purged.[20]

Pulp Mill Wastes

Pulp mill wastes originate from two chemical processes, the sulfite and Kraft processes discussed in Chapter 4. Wastes from these processes are very complex, and those of the Kraft process are slightly more toxic than those from the sulfite process. A composite sample of the effluent from a Kraft pulp mill including waste from the bleach plant can include chlorinated lignins, resins, fatty acid soaps, hydrochloric acid, sulfuric acid, chlorosulfonic acid, and various inorganic salts. Additional products which are fairly rapidly lost from the effluent through volatilization or natural breakdown include sodium sulfide, sodium hydrogen sulfide, methyl mercaptan, hydrogen sulfide, and the sodium salts of resin acids and soaps. These latter substances are the ones to which most of the toxicity is attributed; levels which cause fish mortalities range from 1 milligram per liter to 3 mg/l. Considering the Kraft process waste as a whole, lethal levels can range from 500 to 10,000 parts per million, depending upon the type of receiving water and fish species used in the test.[21] In general, salmon is a cold-water species and is more sensitive to toxic wastes than warm-water species. The BOD in the Kraft process waste is relatively small compared to the sulfite process.

Waste water from the sulfite process, sulfite waste liquor (SWL), has been found to be toxic at concentrations of 1,000 to 20,000 parts per million,[22] a range in toxicity that is not much greater than the range in toxicity for Kraft waste. Although some researchers have concluded that Kraft waste is much more toxic than SWL, a study team in Washington actually found that synthetic and actual Kraft waste was less toxic to salmon than SWL.[23]

Although the toxicity of pulp wastes to fish is low, oysters are approximately one hundred times more sensitive, and the toxic level for SWL to oysters has been placed as low as 16 ppm.[24] The largest problem posed by sulfite process mills is the BOD content of the wastes discharged. Waste with 10 to 14 percent solids has a BOD of about 30,000 mg/l. Because of the extremely high BOD it appears that the biggest threat from sulfite mills is the oxygen consuming demand of their wastes. Studies in Puget Sound, Washington, using the oyster larval bio-assay method[25] indicate that waters in the bays near pulp mills contain levels of waste that are detrimental to the immature oyster. Indeed, when one flies over some of the bays in Puget Sound, it is possible to see large black water plumes emanating from the pulp mill sites. This black water certainly looks toxic enough; however, visual impressions are unreliable since some effluents appear obnoxious but are not toxic at all, while others that are invisible are extremely toxic.

Other Industrial Wastes

Myriad toxic ingredients exist in complex industrial wastes. It would serve little purpose here to sift through all of these various types of waste and indicate their effects on aquatic life. Instead, we will consider selected components from a variety of industrial wastes. These particular examples were chosen more on the basis of frequency of occurrence than for toxicity. Phenol (carbolic acid) is highly soluble in water and organic solvents, and is a product of distillation of wood, gas works, coke ovens, oil refineries, chemical plants, and human and animal wastes. Phenol is lethal to fish, particularly trout, in concentrations from 1 to 2 mg/l.[26] It is somewhat less toxic to microorganisms, and some may actually utilize it as an energy source, thereby rendering it non-toxic in a relatively short time. Phenol is not only lethal to fish; it can also cause taste and odor problems by being absorbed into the flesh of the fish and volatilizing following cooking. This effect is quite often more detrimental than a direct toxic effect since it affects more fish over a greater area. In addition, phenol entering public water supplies causes taste and odor problems for the consumer when these waters are chlorinated.

Another commonly occurring toxicant is chromium, a product of metal plating, aluminum processing, leather tanning, and paint, dye, explosive, and ceramic manufacturing. This metal, as dichromate, has been found to be lethal in concentrations of 10 to 200 parts per million. These are acute lethal levels, i.e., levels at which mortality occurs within a relatively short time. They do not indicate the tolerance levels at which fish can survive indefinitely. Tolerance levels for long-term survival would be much lower in most cases.

Ammonia is another good example because the effect of other water-quality characteristics on its toxicity helps to explain the range in lethal levels expressed for some of the chemicals already discussed. The form in which a chemical exists depends on such water-quality characteristics as pH, temperature, oxygen content, and competing chemical ions. Ammonia has been found to be toxic at about 1 mg/l. However, this toxicity is due to the undissociated NH_3 molecule. If ammonium chloride (NH_4CL) is added to water, the amount of NH_3 molecule liberated from this compound varies tremendously depending upon pH. As pH increases, the amount of undissociated NH_3 increases, usually about ten times for every unit increase in pH. Superimposed upon the variability of water quality is the variability of the organisms involved. The size of the organism as well as the species, the time of year, and the relative state of health can all greatly affect the tolerance level for a toxicant. One can appreciate the difficulty of estimating the effect of a waste containing several ingredients on the biological community in a receiving water. These difficulties are compounded since water quality standards must apply over a broad area encompassing several watersheds with varying water quality characteristics and communities of organisms.

Suspended Solids and Urbanization

Turbidity, or suspended solids, has a very significant effect on aquatic life. Turbidity is not directly harmful to fish except at extremely high concentrations. Many species are able to tolerate suspended inorganic solids in concentrations as high as 100,000 to 200,000 parts per million, although harmful effects are sometimes noted at concentrations around 20,000 ppm. However, suspended solids can have a significant adverse effect below the directly lethal levels. Concentrations as low as 100 to 200 ppm can ultimately reduce aquatic life. In many cases the effects of sediment are more permanent than more immediately toxic industrial wastes, because, once deposited, they are very difficult to remove.

In their natural state, watersheds normally supply nearly sediment-free water to the drainage streams. Runoff is low in sediment content because vegetation holds soil in place and maintains the natural erosion process at a low rate. Removing vegetation from the watershed to make way for housing and commercial development greatly accelerates the erosion process, and sediment reaches streams in amounts much greater than normal. Studies have shown that the sediment yield from urbanized watersheds is five to two hundred times that from rural watersheds.[27] Sediment yield from developed land can be thousands of times that from undeveloped land.

Clearing vegetation from land under development is usually done very carelessly. Developers usually follow the quickest and simplest procedure—complete removal of trees and ground cover—baring the soil to erosive processes. Even though natural vegetation has great appeal to buyers, there seems to be little incentive for the construction industry to preserve it.

What happens to aquatic communities in streams receiving sediment eroded from bared slopes? The productivity and diversity of life in a stream is related to a large extent to the quantity and quality of bottom materials—the rubble and gravel in stream beds. A clean stream bed, relatively free of fine sediment and silt, exposes a large surface area of rocks, which are the habitat of insect nymphs and larvae, the most significant source of food for fish. Clean rocks also provide surface for attachment and growth of microscopic plants, which in turn serve as food for the insects. Trout and salmon require clean stream beds, because during spawning they deposit fertilized eggs in depressions they dig in the bed, and the eggs sift down into crevices among the rocks. The eggs are later covered with gravel by the fish and finally incubate a foot or more below the stream-bed surface. Since the developing eggs require a large oxygen supply and continuous elimination of waste products, adequate water flow through the gravel is necessary to maintain a high oxygen content (seven to eight parts per million) and carry away the damaging waste products. Small solid particles and silt entering the stream will eventually settle to the bottom and sift down among the rocks. If the input is continuous, the spaces among the rocks will fill with sediment, severely reducing the intra-gravel water velocity and the oxygen supply to incubating eggs.

Many investigations have documented various sources of inorganic sediment, including runoff from logged-over forests, discharge of mine tailings, highway construction, and irrigation return water. In some cases the quantity of bottom fauna has been reduced from 70 to 90 percent. Survival of trout eggs to hatching in unaffected areas of one stream was 90 percent, while in areas below the entrance of sediment-laden irrigation return water, survival was only 10 percent.[28] The average sediment concentration of the affected stream was 200 parts per million, which appears only slightly cloudy to the eye—1000 times less than a lethal concentration to adult fish. At higher sediment concentrations (1000 ppm) no survival was observed. Survival, then, is largely related to water flow through the gravel, since it affects oxygen delivery to incubating eggs.

In Thornton Creek, within the city of Seattle, the quantity and quality of biota gradually deteriorate as one progresses downstream. The upper section of the stream, surrounded by a well vegetated watershed, has very clear water with a clean and porous gravel bottom. In this habitat several kinds of relatively sensitive insects and other macroscopic animals exist. Downstream, near the entrance of the stream into Lake Washington, fewer species exist in fewer numbers than upstream. The bottom is much less porous, fine sediment fills crevices among the rocks, and the habitable area is greatly reduced. The kinds and abundance of organisms in the lower stream reflect a mildly polluted condition even though waste effluents, as such, do not exist. Thornton Creek has been degraded by sediment washed off roads and graded slopes, probably accompanied by fertilizer and pesticides leached from residential lawns, as well as oil and organic debris washed from streets.

Water quality is deteriorating in many other streams and rivers throughout the country, in many cases with more dramatic consequences than in Thornton Creek. It is virtually impossible to build houses without having some ecological effect on nearby streams. The complicated nature in which sediment affects aquatic ecosystems precludes establishing a standard concentration level to protect them adequately. The safest method for developing land is to limit the input of sediment to a minimum by employing strict controls. Even though preventing all adverse effects is unlikely, they can surely be minimized. Obviously one of the most effective controls would be to prevent the removal of vegetation that is important in preventing erosion. This vegetation would serve the secondary purpose of preserving the aesthetic appeal of the area. Other controls, such as catch basins and green belts around developments, are no doubt necessary. However, the closer to its source that erosion is controlled, the more effective control will be established.

Heat

Heated water can help enrich as well as inhibit biological production in aquatic systems. Enrichment in the strictest sense refers to an increase in nutrients that are required for biological growth. Although heated water discharged to lakes and streams from power plants favors certain tolerant species, heat is generally detrimental to existing communities of organisms. The most direct effect of heat is mortality of organisms in receiving waters.

The quantities of waste heat produced from thermal power production will increase exponentially in the coming decades. There are many effects of heated water discharged to aquatic environments, including immediate mortality, increased growth of some organisms, increased biological production, changes in community structure, changes in species dominance, lowered oxygen content, and changes in the seasonal temperature regime which inhibits reproduction.

The actual temperatures that are lethal to various aquatic species are a function of test conditions. The length of time that test fish are acclimated to a temperature lower than the test temperature is very important. The lethal temperature can actually be increased as the acclimation temperature is increased. Assuming adequate acclimation to test temperatures has been achieved, lethal limits for temperate fish range from 75° to 105°F. The maximum limit for warm-water fish that is presently being set to avoid mortality is 93°F. The level for trout is 68°F. Of course, these levels only assure protection from direct mortality. Much lower temperatures are necessary to protect reproduction, growth, and other normal activities.

Of all the pollutants we have discussed, heat causes the most subtle changes in the structure of the biological community, because temperature intimately regulates life processes. This is particularly true of aquatic organisms whose body temperature varies with the temperature of the environment. As temperature changes, the activity of aquatic animals is affected. Many of the blue-green algae can tolerate higher temperatures than other algae. For example, some blue-green algae are among the few organisms able to live in hot springs like those in Yellowstone Park, where temperatures reach 158°F, so it is not surprising that blue-green algae are also found dominating the plant communities below thermal waste discharges. Ambient water temperatures of 86°F have been shown experimentally to result in a shift in dominance of biological communities to forms that cause nuisances, e.g., blue-green algae.[29] In general, blue-green algae do not supply food to consumer organisms as do some other algal forms. Furthermore, the slimy mats that they cause are unsightly and create tastes and odors in water supplies. Certain consequences of eutrophication, therefore, can also result from temperature increases.

Productivity does increase with increasing temperature, and slight increases in temperature may even be quite beneficial, if only slight changes in the community structure occur. That is, a change could cause greater production of a desired variety of fish for sportsmen. Temperatures near heated water discharges during the winter are generally closer to the optimum for growth and activity of game fish species than normal river temperatures. As a result, fish feed, grow, and are caught at a greater-than-normal rate in the winter in these areas. In fact, it has been observed that fish populations shift in the direction of game fish species in the vicinity of heated water discharges during the winter.[30] In the Wabash River, Indiana, the kinds and abundance of fish and invertebrate bottom organisms decreased in summer near heated discharge areas, and increased in the fall as temperatures decreased. Because fish are mobile they exercised temperature preferences and vacated affected areas when temperatures reached 95°F. They dispersed throughout the heated areas of the rivers into zones separated by as little as 2 to 3°F. Attraction of the heated area in winter is certainly a beneficial effect. However, whether this phenomenon is generally beneficial, considering the detrimental effects from high temperatures during the summer, has not been adequately determined.

In the Pacific Northwest there is an additional problem related to thermal discharges and mixing zones, because some of the more important species are migratory—salmon and steelhead trout. Although leaving part of the river unaffected by heated water discharges is probably satisfactory for upstream migration, it may deter downstream migration. The young downstream migrants are relatively small and are weak swimmers. Thus, if they are swept into an outfall area, they may not be able to avoid the heated water, which may be lethal. Fish can generally detect and avoid unfavorable water quality conditions, including temperature. However, the high temperatures coupled with the inept swimming ability of the young fish may result in significant losses during downstream migration. Even if fish are not killed outright as they are swept through the area, it has been shown that short exposure to greater-than-lethal temperatures will decrease their resistance to predation.[31]

An even more serious consideration in the protection of migratory fish and weak-swimming invertebrates is their possible entrapment in cooling-water intakes. In traveling through a cooling system even briefly, abrupt temperature changes of as much as 25°F will be encountered. In some instances damage to organisms has been shown from such exposure, but in other cases there has been no apparent effect. In general it seems that fish and most aquatic life can tolerate rapid increases in temperature if these increases do not approach the lethal level and the exposure is not prolonged. However, the longer they are exposed to heat that would ultimately be lethal, the greater is the likelihood of damage.[32]

There will no doubt be increasing damage from heated water discharge as thermal power generation increases. Cooling towers are not practical in every instance and may not even be desirable where fouling occurs or where fog presents a problem. Use of the waste heat for irrigation and fish rearing will probably not contribute much to solution of the heat dispersal problem. The effects of heat can be minimized by proper siting of generating stations, design of intake structures, and discharge diffusers. For these alternatives to be effective, more precise knowledge of the sub-lethal effects and thermal limits for a variety of important organisms and communities must be discovered.

Oil

The source of oil pollution that is most detrimental to the aquatic environment is spillage from tankers. Most of the oil carried in these tankers is crude oil and has quite different characteristics than the refined oil such as diesel or fuel oil. We will consider here the effects of oil spills from the standpoint of toxicity to aquatic life, aesthetic damage, and the very large quantities involved.

Crude oil, such as was spilled in the *Torrey Canyon* and Santa Barbara episodes, is relatively low in toxicity to aquatic life. It is, however, very detrimental to water birds, causing matting of feathers, suffocation, poisoning, and heat loss; large bird kills are common at oil spills. The principal effect on aquatic organisms is a smothering of sessile animals in the intertidal zone. In some local areas the lighter and more soluble components of crude oil (xylene, benzene, toluene) could have a toxic effect. The severe kills of intertidal and subtidal animals in the *Torrey Canyon* disaster and other episodes were actually caused by the detergents used to disperse the oil. Clean-up crews no longer use toxic detergents and now employ absorbent materials to concentrate the oil. If the spill is confined to a small area, substantial losses of intertidal organisms can result, as was the case when 59,000 barrels escaped from the vessel *Tampico* in a small cove in lower California Bay in 1957.

Two other problems of oil spills are the large quantity that is shipped—Alaska's North Slope may produce 100 billion barrels—and the mess that is created when the spill washes ashore. A certain amount of concentration and clean-up is possible, but substantial areas of beaches covered with thick, gooey crude oil will surely result. In confined environments like Puget Sound, in which eddy currents are strong, spilled oil will collect on the beach in a very short time, probably even before it can be intercepted by clean-up crews. Effects may range from large areas becoming unusable for recreation to serious aquatic animal mortalities similar to the *Tampico* incident. At the expected shipping rate from the North Slope (1 million barrels per day), severe degradation of the abundant and aesthetically pleasing shorelines of Puget Sound will no doubt occur even if only a portion of this oil enters by that route. This problem can be most effectively minimized by increasing safety precautions and preventing spills in the first place.

PATTERNS OF COMMUNITY RESPONSE

Sewage, with its effect on the oxygen content of the stream and its intrinsic supply of food material, has selective effects on communities of organisms. In the most severely affected parts of the stream, fewer kinds of organisms are present, in greater numbers, because few can tolerate the low oxygen content. Downstream, algae develop at much higher levels than algae upstream of the introduction of wastes.

Somewhat in contrast to organic wastes, toxic and inorganic sediment wastes generally eliminate most organisms equally, if toxicity is sufficiently high. The effects on biological communities are decreases not only in the kinds of organisms, but also in the numbers of individuals of each species. Typical community changes due to waste additions are summarized in Figure 5–6. However, even with toxic wastes, if the level is low enough to permit the differential tolerance within the biological community to exert itself, species that are more tolerant will usually out-compete less tolerant species, and subtle shifts in the composition of the community will result. When reduction in the number of species simplifies the structure of an ecosystem, species that are less desirable to man generally become dominant. For example, many warm-water fish frequently show more resistance to toxicants than many of the cold-water species. Organic and toxic wastes may enter simultaneously, and interactions occur, with the results shown in Figure 5–6.

Some species important to the diet of desired fish may be eliminated and cause the dominance of less desirable species through competition for available food resources. However, fish are not usually dependent upon a single food item, but rather eat the organisms that are available. Reducing the reproductive success of a favored species via pesticide residues or sealing

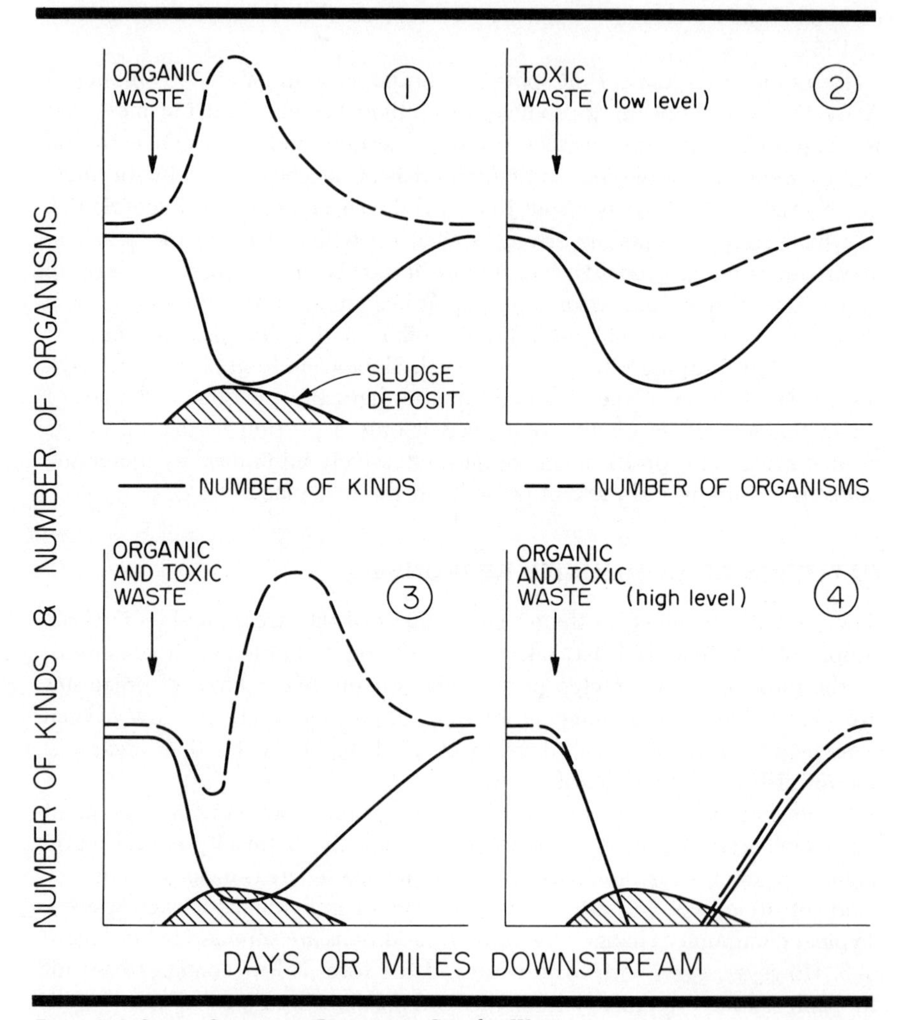

Figure 5-6. Stream Community Response to Complex Waste

(1) organic waste, (2) low level toxic waste, (3) organic and toxic waste combined, and (4) high level toxic and organic waste combined.

From STREAM BIOLOGY FOR ASSESSING SEWAGE TREATMENT PLANT EFFICIENCY, by L. E. Keup. WATER AND SEWAGE WORKS, Vol. 113, 1966. Reprinted by permission of the publisher.

the stream bed with sediment can result in the eventual elimination of the species and cause a shift in relative abundance of the various members of the fish community.

WATER QUALITY STANDARDS

Because of the difficulties in setting standards of water quality, only a few environmental variables important to aquatic life are specifically defined in enforceable standards. These variables are oxygen content, temperature, turbidity, and pH (acidity or alkalinity). To cover all other inhibiting or enriching waste materials, a catch-all phrase like "Nothing can be added that will damage aquatic life" is usually included. Among these are nutrients, pesticides, heavy metals, and complex industrial wastes (organics, gases, oils, and derivatives), as well as physical destruction such as channel alternation and soil erosion. Generalization is extremely difficult with respect to these wastes, and actual or potential damage must be determined and controlled in nearly every case through field and laboratory investigations.

An even broader problem exists with standards for maintaining ecosystem integrity in the face of ever increasing quantities of wastes. Customarily, a particular waste is not considered a pollutant unless it interferes with a use beneficial to man. Some of the more important uses are domestic water supply, irrigation, industrial water supply, recreation, and the support of fish and other aquatic life. A rather extensive gap exists between the criteria of water quality suitable for the first three of these uses and those suitable for fish, other aquatic life, and recreation. With regard to domestic water supply, the recommended levels of copper and zinc for drinking water can be lethal to aquatic life, particularly in soft water. Standards for irrigation supply state that waters containing more than 1000 mg per liter of total dissolved chemicals may be detrimental to crops. Although some organisms can tolerate these levels, an increase of even 100 to 200 mg per liter will produce considerable changes in biological communities.

The recommended water quality standards for various industrial uses, as shown in Table 5–1, are fairly straightforward and appear to be easy to meet. Many synthetic and naturally occurring toxic substances are not men-

TABLE 5–1

Water Quality Standards for Various Uses of Water

	O_2	NH_3	pH	H_2S	Temp. °F	TDS	Cu
Domestic	3	0.5	6–8.5		85	500	1.0
Textile			6–8			150	0.5
Lumber			5–9				
Pulp and Paper			4.6–8.4		85	1000	
Chemical			5.5–9			2500	
Petroleum			6–9			3500	
Primary Materials			3–9		100	1500	
Fish and Other Aquatic Life	5–7	1.5*	6–9	*	48–93	1600	0.1*

Bioassay recommended with water and test species from the area in question because of the great variability of results among receiving waters and organism response.

tioned in the criteria. Water quality characteristics that are important to the survival and growth of aquatic organisms are either not listed (oxygen content) or are considerably outside the limits tolerable for aquatic life. Cooling water, for example, is sometimes discharged at 100°F, a temperature virtually intolerable for all except a small group of aquatic flora. The point here is that, if it were not for the standards demanded for fish, aquatic life, and recreation, our waters would be more polluted than they are now. The criteria for the other beneficial uses could be met much more easily and with less investment in waste treatment than is required to meet the standards set to protect aquatic life.

SUMMARY

Ecosystems that receive an increase in their energy supply or an increase in their ability to utilize available energy supplies are said to be enriched. Materials that commonly cause enrichment of aquatic ecosystems are energy-rich organic wastes and inorganic wastes containing such nutrients as nitrogen, phosphorus, and carbon. The principal effect of enrichment is a shift in community dominance from many types of organisms in low and equal abundance to fewer organisms in relatively high abundance. Frequently, those organisms able to flourish in an enriched ecosystem are nuisances to man. Additional effects of enrichment are the depletion of aquatic oxygen resources and the accumulation of sediment, both of which may shift community dominance.

Enrichment with organic waste can be adequately controlled through treatment, although it is still a serious problem in many parts of the world. Projected increases in the use of agricultural fertilizers and urban development, however, suggest that the control of nitrogen and phosphorus to waters (cultural eutrophication) will be considerably more serious in the years ahead than controlling organic wastes. Phosphorus appears to cause eutrophication in the majority of cases, and some improvement has been evidenced from its control. However, our present knowledge is inadequate to predict how large an increase in nutrients our waters can tolerate without undesirable consequences.

Ecosystem degradation from inhibiting wastes is more apparent to the public because of fish kills that often result from toxicant additions. There are numerous constitutents of industrial wastes that are potentially toxic in receiving waters, although the exact lethal concentrations depend upon the species involved, the type of toxicant, and other water quality characteristics such as acidity, temperature, and oxygen content. Lethal limits may vary from only fractions of a part per million to several thousand parts per million. Because of these variations, broadly applied water quality standards do not usually specify the broad category of toxicants. Instead, toxicity bio-assays are usually required to evaluate individual wastes, and in some cases, the wastes are so complex that their toxic constituents have not been identified.

Low levels of toxicants cause a slow but certain degradation of the community without a noticeable mass mortality, and may be more significant than toxicants that are immediately lethal. Environments and organisms can accumulate toxicants that resist decomposition, such as organo-chlorine pesticides, mercury, and lead, if the intake rate is greater than the loss rate. The subtle effects of accumulated materials can inhibit reproduction or resistance to stress and gradually result in subtle shifts in community dominance. Although suspended solids are not toxic *per se*, their effects on communities are insidious and pervasive through habitat elimination. Such subtle effects are nearly impossible to predict, and the only practical control is to curtail the input of these materials to the environment. Such steps have recently been apparent in the DDT ban, the reduction of mercury output, and the development of non-leaded gasoline.

Heat additions in coolant waters will be a serious future environmental problem because of the increasing demands for electrical energy. Biological communities are very sensitive to temperature, and some change in communities is inevitable as a result of future energy production. Shifts in species distributions near power plants can be expected as a result of even 2 to 3°F increases. However, these effects can probably be greatly minimized by using cooling towers, proper power plant placement, and proper diffuser design. It is vital to avoid subjecting weak-swimming or migratory species to large increases in temperature through entrapment in intake water or through creation of "hot spots" near outfalls.

Wastes are varied in their compositions as well as their real and potential effects on the aquatic environment. What can be concluded about the relative future hazards to the environment posed by these wastes? Pesticides, nutrients, and sediment pose the greatest threat to the future of the aquatic environment. The effects from these wastes are insidious, tend to be long-lasting, and, once they are detected, reversal is usually difficult. The fact that sources of wastes are many and diverse is of the greatest significance. They do not enter the aquatic environment at a single point, and therefore their ultimate control is very difficult. Agricultural and other land development is the biggest source of these wastes. Pressures for increased agricultural production will become increasingly intense in the future, and collection and control of these wastes will be a challenge to environmental engineers.

Oxygen consuming wastes, metals and other inorganics, heat, and toxic synthetic organics probably pose a lesser threat to the environment. For the most part these wastes produce acute but short-term effects. This is particularly true for many inorganic toxicants like ammonia and cyanide, easily decomposed organics like phenol, oxygen-consuming wastes, and heat. If the sources of these wastes are controlled, the local environment should rapidly improve. Control of these wastes is not a complex problem, because they usually originate from factories. Mercury and lead are exceptions, because they have already accumulated to significant levels in the environment and are particularly toxic.

References—Chapter 5

1. McCarty, P. L., *et al.*
"Sources of Nitrogen and Phosphorus in Water Supplies," *Journal of the American Water Works Association*, Vol. 59, March 1967, pp. 344–346.

2. Kuentzel, L. E.
"Bacteria, Carbon Dioxide and Algae Blooms," *Journal of the Water Pollution Control Federation*, Vol. 41, No. 10, 1969, pp. 1737–1747.

3. Kerr, P., and Paris, D. F.
"The Interrelation of Carbon and Phosphorus in Regulating Heterotrophic and Autotrophic Populations in Aquatic Ecosystems," USDI, FWQA, Water Pollution Control Research Series, 16050fgs, July 1970.

4. Hynes, H. B. N., and Greib, B. J.
"Movement of Phosphate and Other Ions From and Through Muds," Resources Board of Canada, *Journal Fisheries Research Board of Canada*, Vol. 27, No. 4, 1970, pp. 653–668.

5. Edmondson, W. T.
"Phosphorus Nitrogen and Algae in Lake Washington after Diversion of Sewage," *Science*, Vol. 169, August 1970, pp. 690–691.

6. Emery, R. M., Welch, E. B., and Moon, C. E.
"Delayed Recovery in a Mesotrophic Lake Following Nutrient Diversion." Water Pollution Controi Federation, Vol. 43, in press, 1973.

7. Middlebrooks, E. J., *et al.*
"Eutrophication of Surface Water—Lake Tahoe," *JWPCF*, Vol. 43, No. 2, 1971, pp. 242-251.

8. Porcella, D. B., McGauhey, P. H., and Dugan, G. L.
"Biological Response to Tertiary Treated Effluent in Indian Creek Reservoir." Presented to the 44th annual Conference of the Water Pollution Control Federation, San Francisco, October 3-8, 1971.

9. Oglesby, R. T.
"Effects of Controlled Nutrient Dilution on the Eutrophication of a Lake in Eutrophication: Causes, Consequences, and Correctives," National Academy of Science, 1969, pp. 483-493.

10. Welch, E. B., Buckley, J. H., and Bush, R. M.
"Dilution as an Algal Bloom Control," *Journal Water Pollution Control Federation*, Vol. 44, No. 12, 1972, pp. 2245-2265.

11. Douderoff, P., and Shumwas, D.
"Dissolved Oxygen Criteria for the Protection of Fish in a Symposium on Water Quality Criteria to Protect Aquatic Life," American Fish Society, Special Publication No. 4, 1967, pp. 13–19.

12. Warren, C. E.
Biology and Water Pollution Control. Philadelphia: W. B. Saunders & Co., 1971, p. 76.

13. Johnson, H.
"The Effects of Endrin on the Reproduction of a Fresh-Water Fish (*Oryzios Latipes*)." Ph.D. Thesis, University of Washington, 1969, p. 136.

14. Reinert, R. E.
"Pesticide Concentrations in Great Lakes Fish," *Pesticide Monitoring Journal*, Vol. 3, No. 4, March 1970, pp. 233–240.

15. *Ibid.*

16. Henderson, C., Ehrglis, A., and Johnson, W. L.
"Organochlorine Insecticide Residues in Fish—Fall 1969 National Pesticide Monitoring Program," *Pesticide Monitoring Journal*, Vol. 5, No. 1, June 1971, pp. 1–11.

17. Johnson, Howard E.
Michigan State University, personal communication.

18. Reinert, *op. cit.*

19. Turney, W. G.
"Mercury Pollution: Michigan's Action Program," *JWPCF*, Vol. 43, No. 7, July 1971, pp. 1427-1438.

20. Gaind, U. S., Langley, D. G., and LeBlanc, P. J.
"Mercury Methylation in Aquatic Environments." Presented to the 44th Meeting of the WPCF, San Francisco, October 3-8, 1971.

21. McKee, J. E., and Wolf, H. W.
"California Water Quality Criteria." Sacramento: California Water Resources Board. 2nd ed. Publication No. 3-A, 1963.

22. *Ibid.*

23. "State of Washington Toxic Effects of Organic and Inorganic Pollutants on Young Salmon and Trout." Dept. of Fisheries Resource Bulletin No. 5, September 1960.

24. Gunter, G., and McKee, J.
"On Oysters and Sulfite Waste Liquor." Report to the Pollution Control Commission, State of Washington, 1960.

25. Woelke, C. E.
"Measurement of Water Quality With the Pacific Oyster Embryo Bioassay, Water Quality Criteria." ASTM STP 416, American Society of Testing Materials, 1967, p. 112.

26. McKee, *op. cit.*

27. Sylvester, R. O.
"Urbanization and Its Impact on Water Resources." Presented to the 42nd meeting of the Water Pollution Control Federation, 1969.

28. Peters, J. C.
"Effects on a Trout Stream of Sediment from Agriculture Practices." *Journal of Wildlife Management*, Vol. 31, No. 4, 1967, pp. 805–812.

29. Coutant, C. C.
"Alteration of the Community Structure of Periphyton by Heated Effluents." Presented to the Pacific Northwest Pollution Control Association, Portland, Oregon, 1966.

30. Data from Tennessee Valley Authority, 1967.

31. Coutant, C. C.
"Temperature, Reproduction and Behavior," *Chesapeake Science*, Vol. 10, Nos. 3 and 4, September–December 1969, pp. 261–274.

33. Welch, E. B., and Wojtalik, T. A.
"Some Effects of Increased Water Temperature on Aquatic Life." Tennessee Valley Authority, 1968.

6 Effects of Wastes on the Human Community

DISEASE

Microorganisms are extremely abundant in the natural environment. Their life history and persistence in particular environments are such that under favorable conditions they can increase rapidly from nearly undetectable numbers to very large populations. Most of the microorganisms that occur in natural environments are harmless to man, and many are important in the breakdown of all organic material. There are fewer organisms that are pathogenic to man and animals. Water has been recognized for over a century as an effective transferring mechanism for pathogenic organisms, but for much of man's existence, epidemics occurred without knowledge of the existence of pathogens. It was not until the middle of the nineteenth century that a firm connection was established between disease transmittance in water and fecal contamination from human population.

Many of the epidemics and plagues of the middle ages were most likely caused by water-borne pathogens, although they were then thought of as acts of God. Fecal contamination and organic pollution of rivers in England, particularly in London during the nineteenth century, resulted in frequent, serious cholera epidemics, and the odors from putrefying bodies were unbearable. Finally, steps were taken to investigate these diseases and determine their causes. Eventually water treatment procedures were developed, and the epidemics were curtailed.

Some of the most important bacterial diseases transmitted by water are poliomyelitis, cholera, typhoid fever, paratyphoid, and bacillary dysentery. Other pathogenic microorganisms found in sewage are not important in terms of epidemics; among these are the organisms responsible for anthrax, brucellosis, and tuberculosis, and amoebas that cause dysentery, tapeworms, and nematode worms. The virus that causes infectious hepatitis is also transmitted in water.

At the present time these diseases are fairly well controlled because of adequate water treatment methods. Treatment includes chlorination, which kills nearly all organisms. It is still a routine practice to test drinking water supplies for their bacterial content as an indication of fecal contamination, because there are still many sources of untreated sewage. The number of coliform organisms in water has been used as an indicator of fecal contamination in water supplies since about 1880. Since it would be difficult because of their very small populations to determine the numbers of pathogenic organisms actually occurring in water even if fecal contamination is present, scientists test for the presence of *Escherichia coli*, an organism that is more abundant and is associated with fecal waste. This organism is a member of the coliform group found on vegetation, in the soil, in the intestines of warm-blooded animals, and other places. *Escherichia coli* lives in the intestines of warm-blooded animals, including man, but is not a pathogen. Therefore, a test for coliform organisms and a determination of the relative numbers of coliforms present is only an indication that fecal contamination of water might have occurred. Further tests can be performed by which positive identification of fecal organisms is possible. However, to conserve time the test is usually only carried through its initial stages. The maximum permissible concentration for pre-treated water is 1 million coliform organisms per milliliter, or if separation of fecal coliforms is performed the limit is 200,000 organisms per milliliter. Water containing more than these levels needs corrective measures if the source is to be used for drinking.

Intestinal bacteria do not persist for long in natural environments. Water is a hostile environment compared to the intestines of warm-blooded animals. Therefore, very near the sewage treatment plant one would expect to find the highest concentration of coliform organisms. Farther downstream, a continual reduction in the concentration of these organisms would occur as die-off commences, as shown in Figure 6–1. The validity of coliform testing as an indicator of pathogenic contamination depends on the fact that coliforms die away more slowly than pathogens. In certain instances, however, coliforms have been found to die faster than pathogens, raising a question

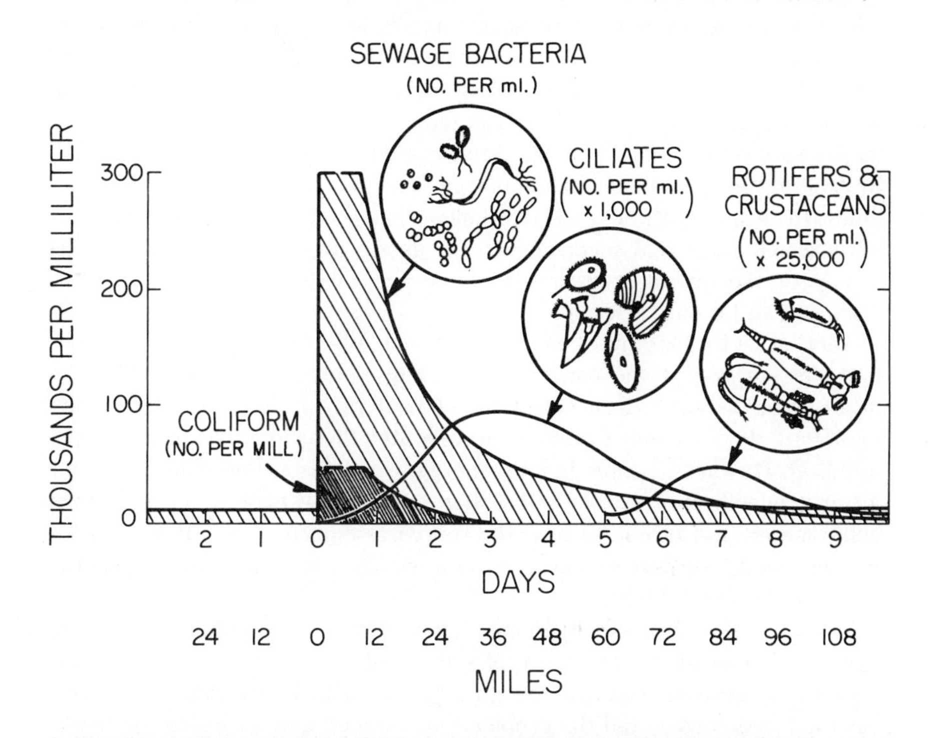

Figure 6–1. Effects of Domestic Waste on Downstream Succession of Microorganisms
Bacteria thrive and finally become prey of the ciliates, which in turn are food for the rotifers and crustaceans.

From STREAM LIFE AND THE POLLUTION ENVIRONMENT, by A. F. Bartsch and W. Ingram. PUBLIC WORKS, Vol. 90, 1959. Reprinted by permission of the publisher.

about the usefulness of coliforms as indicators of pathogen content. When water is chlorinated to control pathogenic organisms, all bacteria are killed, including the saprophytic forms that aid in the self-purification of streams. The self-purification of chlorinated effluents is therefore retarded until new bacterial populations develop when the chlorine content has dropped below the lethal limit.

Air currents also serve as agents for transmittance of bacteria and their reproductive spores. Not as much is known about microorganisms in the atmosphere, with the possible exception of transmittance in hospitals. Man also emits pathogenic bacteria to the atmosphere. Diseases that are known to be transmitted from man to man via bacteria in the atmosphere are diphtheria, septic sore throat, scarlet fever, rheumatic fever, tuberculosis, pneumonia, meningitis, and whooping cough. Contagious diseases caused by viruses are smallpox, chickenpox, measles, German measles, mumps, influenza, the common cold, and psittacosis.

The distance bacteria can travel in air and still be effective upon reaching a host varies, depending upon the nature of the particles that carry them, and also on the species of bacteria involved. Bacteria cannot grow on substrates in the atmosphere but they can be protected depending on the kinds of materials they are attached to. The reproductive spores of many microorganisms are generally more resistant to environmental pressures than bacteria. In general, the persistence of microorganisms may vary from none to many miles. In hospitals the effective distance of pathogens has been defined in fairly exact terms and in many cases beds are arranged far enough apart to impede the spread of disease from one patient to another. This effective distance can be only a few feet.

Another factor that must be considered when discussing pathogen transmittance by air is that the concentration of microorganisms is much smaller in air than in water. Concentrations of bacteria in clean air may be less than 1 to 10 per cubic meter, while in water there may be thousands per milliliter. However, people breathe more air in a day's time than they drink water. Concentrations in the atmosphere of hospitals may go up to 100 per cubic meter, and of course near the source of emission such as around the person's mouth or near an outfall from a sewage treatment plant concentrations can be much higher.

What are the chances for people, exposed to an atmosphere for various periods of time in the proximity of sources of microorganisms, to contract diseases transmitted through the air? As yet, sufficiently intensive studies have not been made, and the problems need much greater resolution before effective control can be considered.

The land also includes important vehicles for disease organisms. Certainly garbage, improperly handled and concentrated in one area, could provide an important food source for large populations of rats. Rats can be hosts for fleas whose bite causes the plague so well known in man's history. Solid wastes, principally paper, that are normally disposed of in landfill areas may not be direct sources of disease-causing organisms if fecal contamination is not present. Landfill areas would probably represent a reservoir of disease only insofar as food supplies are available to produce organisms able to transport diseases from other sources of contamination. Also, these food supplies could serve to produce huge populations of nuisance and disease-carrying organisms such as mosquitos, various kinds of flies, rats, and birds. If fecal wastes are present, direct transmission of diseases to people could occur.

There are effects of waste production which are undoubtedly important but are not precisely known. As urban centers increase in population density, the space for each individual decreases and the sights that he sees every day become more and more drab. There is no doubt that these influences take some toll on man's mental condition. As the space we can occupy becomes smaller, traffic snarls become more severe, traveling times within the city become greater, noise levels rise, and anxiety and tension increase. This could well lower irritability thresholds, and we could gradually become more aggressive in response to external stimuli. To what extent this is presently occurring and what the future holds in response to further increases in density is unclear. Everyone can judge what aesthetically pleasing experiences are, and give them whatever relative importance he wishes. Man seems to prefer the warm (85°F), humid climate, with green areas of grass with scattered trees. This environment is typical of the African savannah where he probably originated.[1]

CHEMICAL TOXICANTS

Toxicants in Water and Food

Water is an important carrier of toxicants and humans are very vulnerable because of their extensive use of water. As a result, there are drinking water standards that designate allowable limits for the concentration of various toxic elements in water supplies. Toxicants which have standards ascribed to them are arsenic, cadmium, chromium, copper, boron, iron, nitrate, zinc, cyanide, pesticides, detergents, and radioactivity. The permissible concentrations of these chemicals in public water supplies are shown in Table 6–1. Although these standards are expressed in concentrations, it is the absolute

TABLE 6–1

Selected Surface Water Criteria for Public Water Supplies

CONSTITUENT	PERMISSIBLE CRITERIA
	(mg/1)
Arsenic	0.05
Boron	1.0
Cadmium	0.01
Chromium	0.05
Copper	1.0
Cyanide	0.20
Iron	0.30
Nitrate (and nitrites)	10
Zinc	5
Detergents (Methylene blue, active substances)	0.5
Aldrin	0.017
Chlordane	0.003
DDT	0.042
Dieldrin	0.017
Endrin	0.001
Heptachlor	0.018
Heptachlor epoxide	0.018
Lindane	0.056
Methoxychlor	0.035
Organic phosphates plus carbamates	0.1
Toxaphene	0.005
	(pc/1)
Gross beta	1,000
Radium-226	3
Strontium 90	10

From Report of the Committee on Water Quality Criteria. Federal Water Pollution Control Administration, U.S. Department of Interior (1968).

quantity received at any given time that determines the chemical's effect. In fact, many of these standards were developed through knowledge of the effect of a given quantity and the time-rate of water intake.

The effects of most of these chemicals are not cumulative, but rather are dependent on a given dose. That is, a threshold level exists, and if the intake is below that level, the water supply is safe. For example, about 1 mg per liter of fluoride ion is helpful in preventing dental caries. However, at concentrations above 1.5 mg per liter fluoride can cause teeth mottling and fluorosis. Some water contaminants are cumulative; pesticides and some radionuclides are typical examples. Excretion rates are very low, and buildup occurs from small intakes. The recommended standards for chlorinated hydrocarbon insecticides range from less than 0.1 mg per liter to about 0.001 mg per liter. However, there is the stipulation in drinking water standards that these concentrations only refer to the intake via drinking water and do not take into account other sources, such as food. Estimated intake of pesticides from food in the United States is shown in Figure 6–2, based on 30 food store samples from 27 cities. DDT comprises the bulk of chlorinated organic pesticides, which in turn account for about two thirds of all pesticide consumption. The measured levels of residues are well below tolerance limits set for food by FDA and the World Health Organization. Given these ingestion rates it would take 8 years to reach a concentration of 5 ppm in humans if there was no excretion. Disregarding food sources, there are still many unknowns about the degree of accumulation in human body tissue as a result of exposure to very minute concentrations of stable pesticides in drinking water supplies. As we have discussed in earlier sections, pesticides build up to fairly high concentrations in the organs and particularly the fat of many animals in the food web, and the resulting residues can influence the reproductive rates of these animals. Unfortunately, information about similar effects in humans is limited. From all indications hard pesticides, such as DDT, are not harmful to man at ordinary environmental concentrations. The significant decrease in the intake of chlorinated hydrocarbon pesticides, principally DDT and its analogs, from 1966 to 1970 is most encouraging and illustrates increased control measures, as shown in Figure 6–2.

Certain forms of radioactivity can also accumulate in the body. Those nuclides that are particularly important are the ones whose stable element has some metabolic use. Strontium 90 is a nuclide of great significance because it can replace calcium in bone tissue, and since it has a half-life of 28 years, radioactivity from this source can accumulate in the body.

The accumulation and effect of an insecticide or a radioactive nuclide depend on its physical or biological half-life. An insecticide does not decay in the same way that a radionuclide does. Rather, it is metabolized and broken down into non-toxic products either by soil microorganisms or within an organism. Furthermore, some compounds are metabolized at much different rates than others. The half-life of DDT in the environment, for example, is about 15 to 20 years. The herbicides 2,4,5–T and 2,4–D decompose

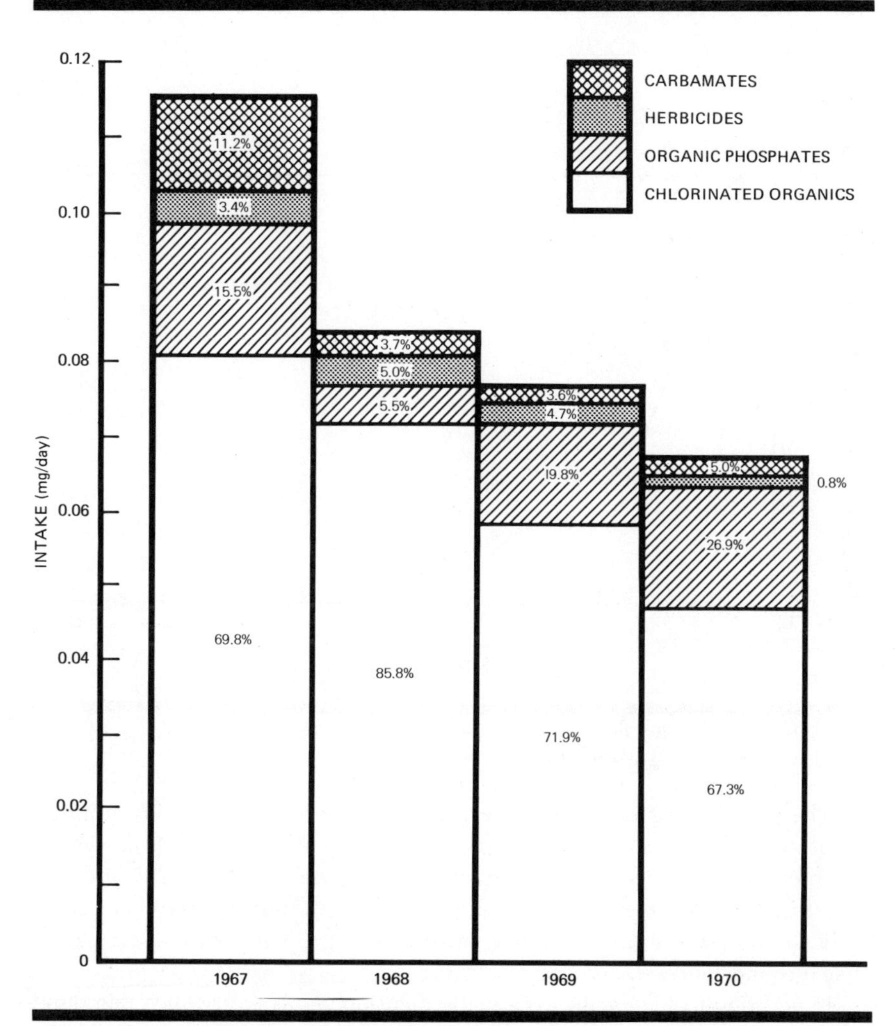

Figure 6–2. Distribution of Total Daily Human Intake of Organic Chemical Pesticides in the United States, 1967 to 1970

Note the major contribution of chlorinated organics and the significant decrease in that component.

From DIETARY INTAKE OF PESTICIDE CHEMICALS IN THE UNITED STATES (III), by R. E. Duggan & P. E. Cornelliussen. PESTICIDES MONITORING JOURNAL, Vol. 5, 1972.

through microbial action at much different rates, with 2,4,5–T being much more stable. See Figure 6–3. Biological half-life, on the other hand, refers to the rate at which the organism discards the particular chemical. Some chemicals are excreted by the body relatively rapidly, so there is little build-up. Others such as chlorinated hydrocarbons, insecticides which are highly soluble in fat, can build up to extremely high concentrations. These concentrations can be metabolized under stress conditions when the animal calls on its fat storage.

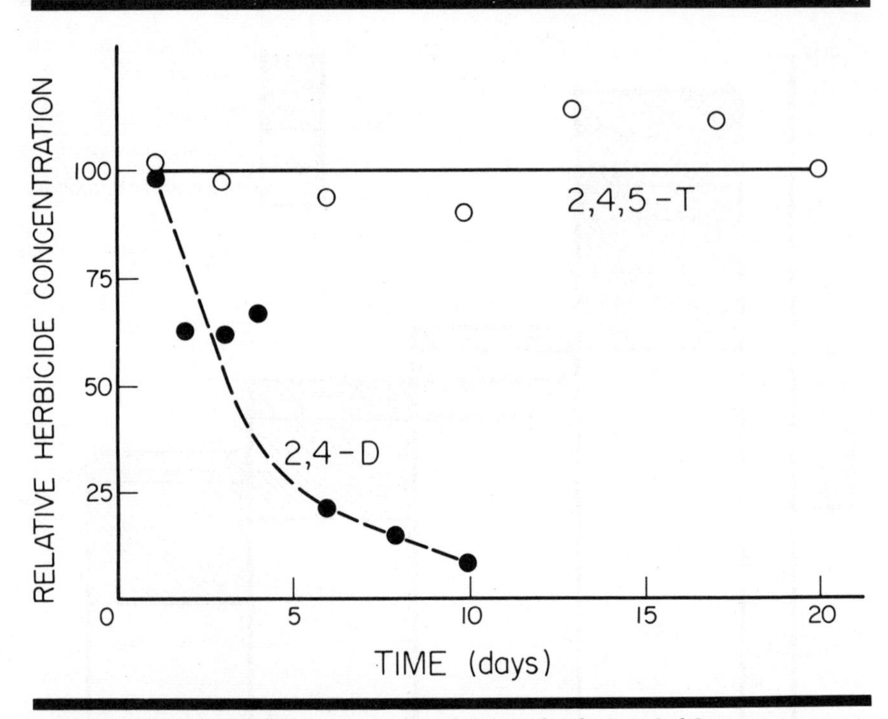

Figure 6–3. Rate of Microbial Decomposition of Two Herbicides in a Soil Suspension
By J. S. Whiteside and M. Alexander, in WEEDS, Vol. 8, 1960.

The safe limit of radioactivity exposure per person is about 170 millirads (units of whole body radiation) per year. By way of comparison, we are exposed naturally to 52 millirads per year at sea level and 207 millirads per year at 5000 feet elevation from cosmic rays and naturally occurring radioactive potassium-40. Fallout, on the average, is less than these natural background sources.

In addition to the chemicals already mentioned, many organic compounds that enter water supplies every day are not being monitored or even being considered as potential hazards. One obvious reason for this lack of controls is that there are so many compounds involved; one really doesn't know which ones are significant. Work is needed to identify the chemical nature of these trace organic contaminants and their effects on organisms. Without precise cause and effect relationships, effective control procedures will probably not be developed.

Airborne Toxicants

Effects of airborne toxicants on human populations can best be illustrated by three outstanding examples in which mortalities have resulted from extreme air pollution conditions. In all of these cases—the Meuse Valley, Belgium, in 1930; Donora, Pennsylvania, in 1948; and London, England, in 1952—high pressure weather systems coupled with temperature inversions occurred during the air pollution episodes. Temperature inversions reduce the quantity of air available to dilute pollution and therefore concentrate the wastes. At Meuse, a dense fog covered the area for a week. This valley had several factories, coke ovens, blast furnaces, steel mills, power plants, glass factories, lime furnaces, zinc reduction plants, a sulfuric acid plant, and a fertilizer plant. After the first three days of temperature inversion and the onset of severe air pollution, many people complained of respiratory difficulties. In total, 60 people died, which was 10 times the normal mortality rate for that period, and thousands were ill. The respiratory complaints included throat irritation, hoarseness, cough, shortness of breath, constriction of the chest, nausea, and vomiting. The consensus of professional opinion was that sulfur dioxide and sulfur trioxide aerosols were the chief causes of respiratory difficulties and mortalities.

At Donora, a temperature inversion and anti-cyclonic weather system persisted for four days. During this time, 17 deaths occurred, compared to a normal mortality rate of two. Respiratory irritations affected 5,910 persons, or 43 percent of the population. The symptoms were similar to those at Meuse, including coughs, chest constriction, headaches, painful breathing, nausea, vomiting, and excessive nasal discharge. Sulfur dioxide, which reached concentrations estimated at 0.5 to 2 ppm, and the other oxidation products of sulfur were considered to be responsible. The 1952 episode in London claimed 4,000 excess deaths due to the extreme air pollution. The same respiratory irritations were reported as in the other two examples, and again the consensus of opinion was that sulfur dioxide accounted for most of the problem. Concentrations ranged from 0.7 to 1.3 ppm, six times the usual level for this chemical. Other episodes have occurred in London. In 1948, 300 excess deaths occurred as a result of a severe temperature inversion, and 1,000 excess deaths occurred again in 1956. All of these situations occurred in areas of intense industrialization in which coal was a main source of fuel. Of course, sulfur dioxide (SO_2) is an important waste product from the burning of coal and may explain part of the emphasis on this chemical in defining the cause of the problem. Some investigators felt that the effect was not due solely to sulfur dioxide, since this gas may operate synergistically with aerosol particles. Experiments have shown that test animals show no effect from concentrations of SO_2 much higher than those associated with the episodes just described. However, effects are pronounced if SO_2 is combined with aerosol particles.[2]

A possible explanation for this synergistic effect is that gases alone (SO_2 included) usually do not penetrate deeply into the bronchioles of the human lung system. With aerosol particles, a surface is provided for absorbing toxic compounds, and deeper penetration into the lung system is possible. Sensitive membranes deep in the lungs then become irritated. A time-concentration relationship necessary to produce certain effects is shown in Figure 6–4.

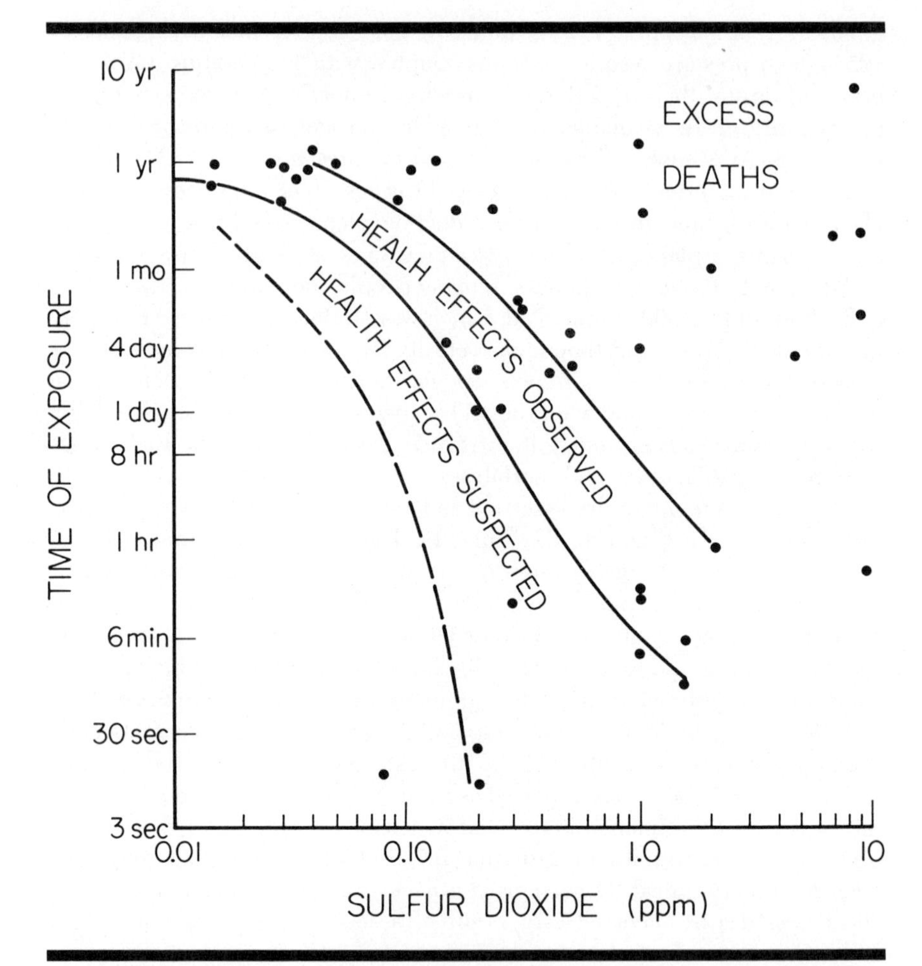

Figure 6–4. Health Effects Resulting From Time-Concentration Exposure of Man or Test Animals to Sulfur Dioxide.

From AIR QUALITY CRITERIA, U.S. Public Health Service, HEW, March 1967.

Some other important toxic chemicals found in air include chromium, beryllium, manganese, ozone, and carbon monoxide. We will briefly describe some of the known effects on humans exposed to various levels of these chemicals to provide a comparison of their potential damages. In one carefully observed situation, an individual was exposed to low concentrations of chromium in the atmosphere for a period of 10 years. After 9 years the individual died, and autopsy revealed concentrations up to 150 micrograms per gram in the lungs and lesser amounts distributed throughout the body. Thus the individual accumulated chromium to fairly high levels. Incidentally, chromium concentrations can be found in the atmosphere above most cities in the United States at levels above those considered to be an effective threshold, as far as human health is concerned. Studies on the effects of beryllium have shown that concentrations of 0.01 to 0.1 microgram per cubic meter cause chronic effects, including fibrosis of the lungs.

The inhalation rate for man is about 4 to 6 liters of air per minute at rest, and this rate increases 5 to 10 times following exertion. The content of lead, copper, iron, and zinc over many United States cities ranges from 17 to 50 micrograms per cubic meter. Applying this inhalation rate to such ambient concentrations provides an estimate of about 1 microgram entering the lungs per minute, or about 1.5 mg per day. The effects of such an intake over a long period of time are difficult to determine. However, it is important to recognize that although water in which toxic chemicals exceed recommended standards can be avoided by using an alternative safe supply, it is much more difficult to avoid contaminated air.

The biological effects of the gases mentioned above are known. As mentioned in the discussion of the air pollution episodes, sulfur dioxide causes restricted breathing, irritation of the respiratory tract, and resistance to pulmonary air flow at concentrations of 2.5 to 8 ppm. The effect is greatest on those who already suffer from such afflictions—air contaminants aggravate respiratory problems. Ozone also causes symptoms of irritation in the respiratory tract, dryness of throat, and headache at concentrations from 0.5 to 1 ppm. This gas frequently is present during air pollution episodes. At 3 ppm, a mild pulmonary edema develops; ozone is lethal to rats after 240 hours at this concentration. At 4 ppm, pulmonary edema is severe, and mortality results.

Carbon monoxide (CO) has an affinity for hemoglobin in the blood, about 200 times that of oxygen. It forms carboxy hemoglobin (COHB), reducing the blood's ability to carry oxygen to tissues in the body. The amount of carboxy hemoglobin (COHB) formed depends on the CO concentration and the exposure time. At 1 percent COHB in the blood, no effect has been noted regardless of exposure time. At 2.5 to 3 percent COHB, time interval discrimination and visual acuity are impaired. At times, the concentration of CO in many United States cities reaches 50 ppm, and during rush hours in Manhattan concentrations of CO can be as high as 15 ppm for as long as 8 hours. Under these conditions it would be possible for COHB to reach levels of 2.5 percent in the blood. One must distinguish between smokers and non-smokers, however, since smokers can have blood concentrations of 3 to 6 percent COHB. Carbon monoxide is not thought to have a chronic effect, that is, an effect from low concentrations over a long period of time.

Some of the diseases caused by chemical irritants in the atmosphere are laryngitis, bronchitis, emphysema, and cancer of the lung, nasal cavity, and larynx. The frequency of emphysema has been shown to be greater in metropolitan areas than in adjacent non-metropolitan areas. This would tend to incriminate air pollution due to toxic dust, fumes, or gases. Chronic bronchitis is also attributed to the same atmospheric conditions. Exposure to Los Angeles smog is supposed to elicit spasms, edema, and secretions that interfere with breathing processes. Thus, there is no doubt that toxicants in air are irritants at times if concentrations are high enough, and have been known to be sufficiently concentrated to cause death.

It is certain that concentrations known to affect health presently occur in United States cities. The average annual concentration of SO_2 in the United States cities ranges from 0 to 0.16 ppm. The maximum daily average is 0.6 ppm, and the maximum hourly average is 1.3 ppm. The time concentration relationship in Figure 6–4 suggests that these maximum conditions are easily within the region in which serious health effects have been observed. The extent to which man can adapt to these higher levels without severe residual effects, or the extent to which other little-known contaminants are having long-term chronic effects, seems to be an open question at this time.

ATMOSPHERIC EFFECTS

Man has caused other effects on his environment which have indirect biological implications. For example, he adds substantially to the CO_2 content of the air on a global basis by burning vast amounts of fossil fuel. Fossil fuels include all forms of fossilized carbonaceous materials such as coal, oil, and natural gas. Figure 6–5 shows the result of CO_2 production measured in Hawaii over more than a decade. Besides long-term trends, the annual effects

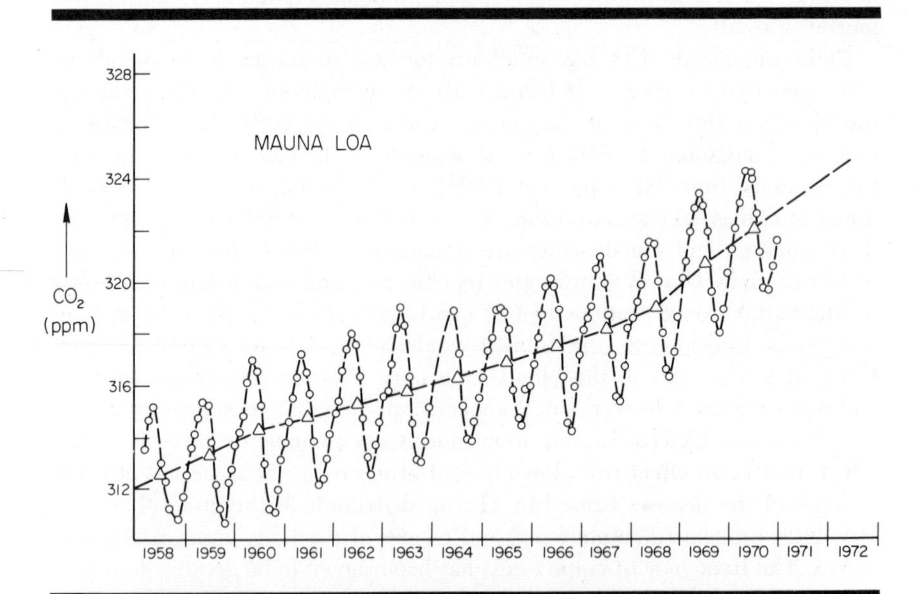

Figure 6–5. Increase in Atmospheric Carbon Dioxide in Hawaii

From THE CONCENTRATION OF ATMOSPHERIC CARBON DIOXIDE IN HAWAII, by J. C. Pales and C. D. Keeling. JOURNAL OF GEOPHYSICAL RESEARCH, Vol. 70, 1965.

of human activity (home heating) and biological activity (photosynthesis) are reflected in an annual cycle with a maximum (for the Northern Hemisphere) in the spring and a minimum in the autumn.

The CO_2 increase is a well documented example of the effect of human activity on the composition of the air. However, the effects of this increase are extremely difficult to predict. Many theoreticians have forecast a variety of climatic effects due to the absorption of infrared radiation by CO_2. A global surface warming of perhaps 0.5°C by the year 2000 is a possibility suggested by some. Others argue that the system is too complex to permit such a simple forecast and point out that there are compensatory mechanisms that could have a cooling rather than a warming effect, especially in the presence of pollutants other than CO_2. Man is tinkering with the composition of the whole atmosphere—Figure 6–5 speaks for itself—without a secure knowledge of the consequences.

We must remember that the effects of changing climate or increased CO_2 are not necessarily bad. A warmer earth *might* be conducive to more food production. Carbon dioxide is known to promote the growth of many plants and food crops. Perhaps man has already augmented his food supply with this ethereal fertilizer. However, we must also recall that climatic changes throughout the earth's history have often had disastrous effects on man and on the biosphere. Besides calamitous events such as ice ages, there have been smaller, more marginal changes that affected life. Advancing glaciers in cool climates, the dying out of Viking farmers in agriculturally marginal lands due to crop failures caused by cooling weather in Iceland and Greenland, and other examples point out the realities of climatic effects.

In order to focus more clearly on these physical effects, which include more than just the CO_2 problem, it is useful to list several areas in which potential problems exist.

Precipitation Modification

Both the *amount* and *chemical composition* of precipitation are inadvertently modified by human activity. Hobbs, Radke, and Shumway[3] showed that there is a systematic increase of precipitation downwind of certain types of paper pulp mills in the state of Washington, and studies in other locations suggest that similar effects exist elsewhere. An interesting point is that the increase *may* be matched by a decrease still further downwind of the source of the modification. This seems reasonable if there is a constant amount of precipitation available in the hydrologic cycle. Man sometimes attempts to modify weather and precipitation amounts by cloud seeding. These efforts are still experimental but have had qualified success in increasing precipitation as well as in suppressing hail.

The chemical composition of precipitation can also be modified by human activity, sometimes in extensive and complex ways. Perhaps the best example has been documented in Scandinavia, where the acidity of precipitation has been increasing steadily for the past decade. Figure 6–6 shows a map of northern Europe with isolines of pH in precipitation which show the extent of this problem. It should be noted that rain is normally acidic with a pH of about 5.5 due to the presence of CO_2, so that the changes in the figure amount to one or a little more pH units (tenfold increase in hydrogen ion concentration).

Thermal Pollution in the Atmosphere

Cities are warmer than the surrounding countryside, and as a result have a different climate, winds, precipitation, and so forth. It is one of the most

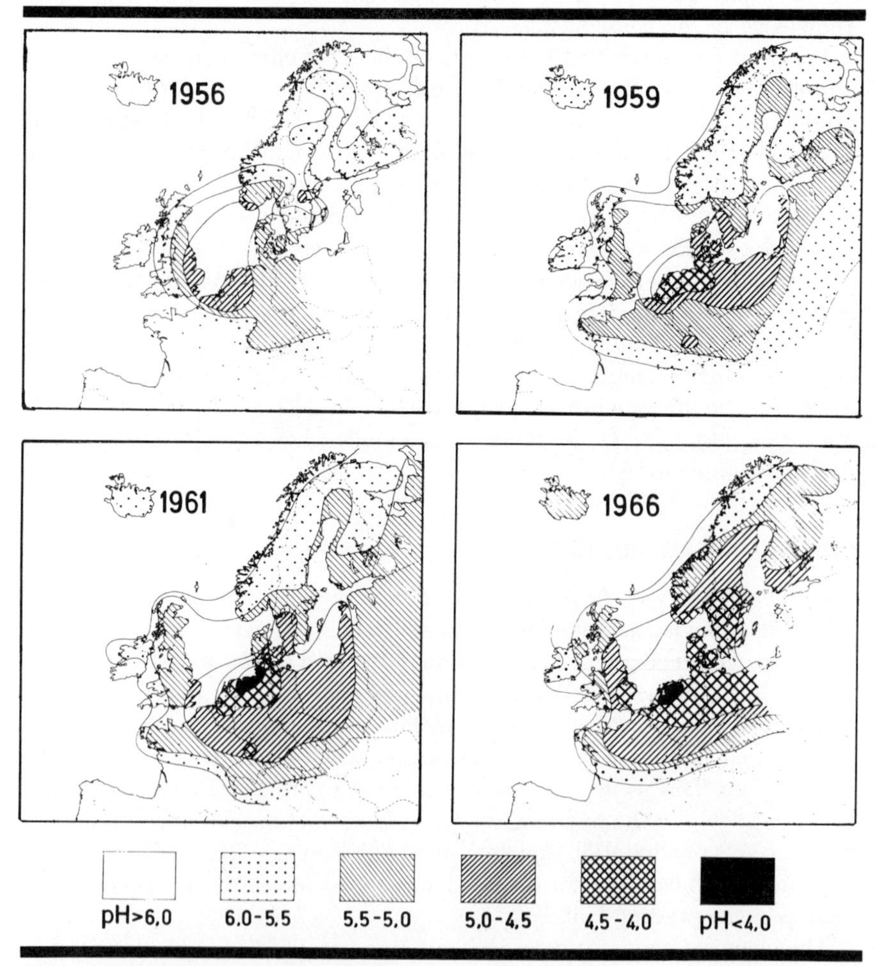

Figure 6–6. Acid Rain in Northern Europe
Gain in average pH value of rainsoaked ground.
From REGIONALA ASPECTER PA MILJOSTORNINGAR, by S. Oden. VANN, Vol. 3, 1969.

dramatic effects of the "heat island" of a city that it often determines whether precipitation occurs as snow or as rain. It snows less in cities than in the surrounding countryside.

Little is known about the overall effect on climate of heat release on a larger scale. However, it seems clear that if present trends continue some effects will be felt. Since most human activity and heat production occur in a small percentage of the area of the earth given to cities, the effects should be noticeable there first. Figure 6–7 shows a projection of the total heat production of mankind as a percentage of the solar input to earth. Since the man-affected part of earth is small, the extent of effects is expected to be more extreme in those areas by a large factor. Hence, heat production already is a recognizable percentage of the sun's heat in urban areas such as the east coast of the United States. The importance of heat production is currently the subject of considerable study by meteorologists and climatologists.

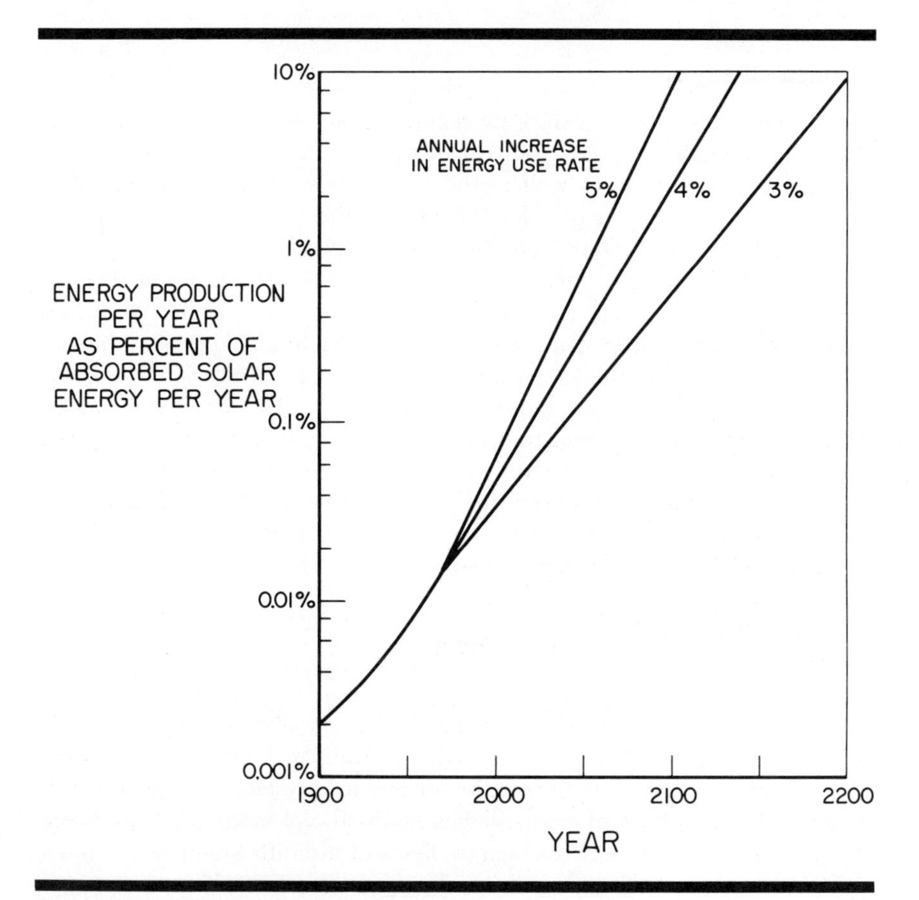

Figure 6–7. Man's Projected Heat Production

Reprinted by permission of A. Hertzberg, Director, Aerospace Research Laboratory, University of Washington.

Albedo Change

Albedo is a term used in meteorology to describe the fraction of solar energy that is reflected away from the earth. It is controlled naturally by the ground, its coverings of plants, the sea, and the cloud cover. Man contributes to albedo by putting particulate matter into the air, by cloud modification, and by surface changes. It is nearly impossible to forecast the effect of the injection of particulate matter on albedo or on cloudiness. Some theoreticians have produced models which predict cooling, leading to an ice age, caused by the reflection of energy by smoke and dust from human activities on a global scale. Others point out that the materials produced (soot, iron oxide, etc.) absorb light and forecast a warming trend. Global data indicate that the amounts of dust in the air are naturally large and variable, which complicates measurement. So far, no trend analogous to the CO_2 trend has been noted.[4] Measurement programs are currently underway to establish any trends that might develop.

Surface Change

Man changes the earth's surface by deforestation, cultivation, urbanization, damming rivers, and spilling oil. All of these change the albedo and heat balance on the surface, as well as the evaporative character of ground or ocean surfaces. Man also produces changes in the roughness of the surface by agricultural and building activities which may affect wind patterns.

The most alarming surface changes are those which might modify the polar snow and ice cover extensively. An oil spill in the Arctic Ocean ice might change the albedo in an important way. Sending ice-breaking tankers or other ships through the ice pack might affect ocean circulation. Damming of the Bering Straits could cause sea ice to melt. The subsequent drastic increase in evaporation from the Arctic Ocean could substantially increase precipitation over all the nearby continental areas (Canada, Siberia, and Scandinavia). Extensive studies of the Arctic Ocean and its ice are presently underway on an international basis, involving the United States, the U.S.S.R., Canada, and the Scandinavian countries.

Contrails and Other Aircraft Effects

High flying jet aircraft frequently leave persistent contrails that contribute to the albedo. Little is known about their effect on albedo, but it is thought to be small. Besides affecting albedo, the contrails may contribute to lower atmospheric weather by sedimentation of tiny ice crystals into lower clouds. This amounts to a form of cloud seeding and is also of unknown importance.

No summary of the environmental effects of aircraft would be complete without mention of the SST and its effects on the stratosphere. Figure 6–8 shows the average temperature of the atmosphere as a function of height.

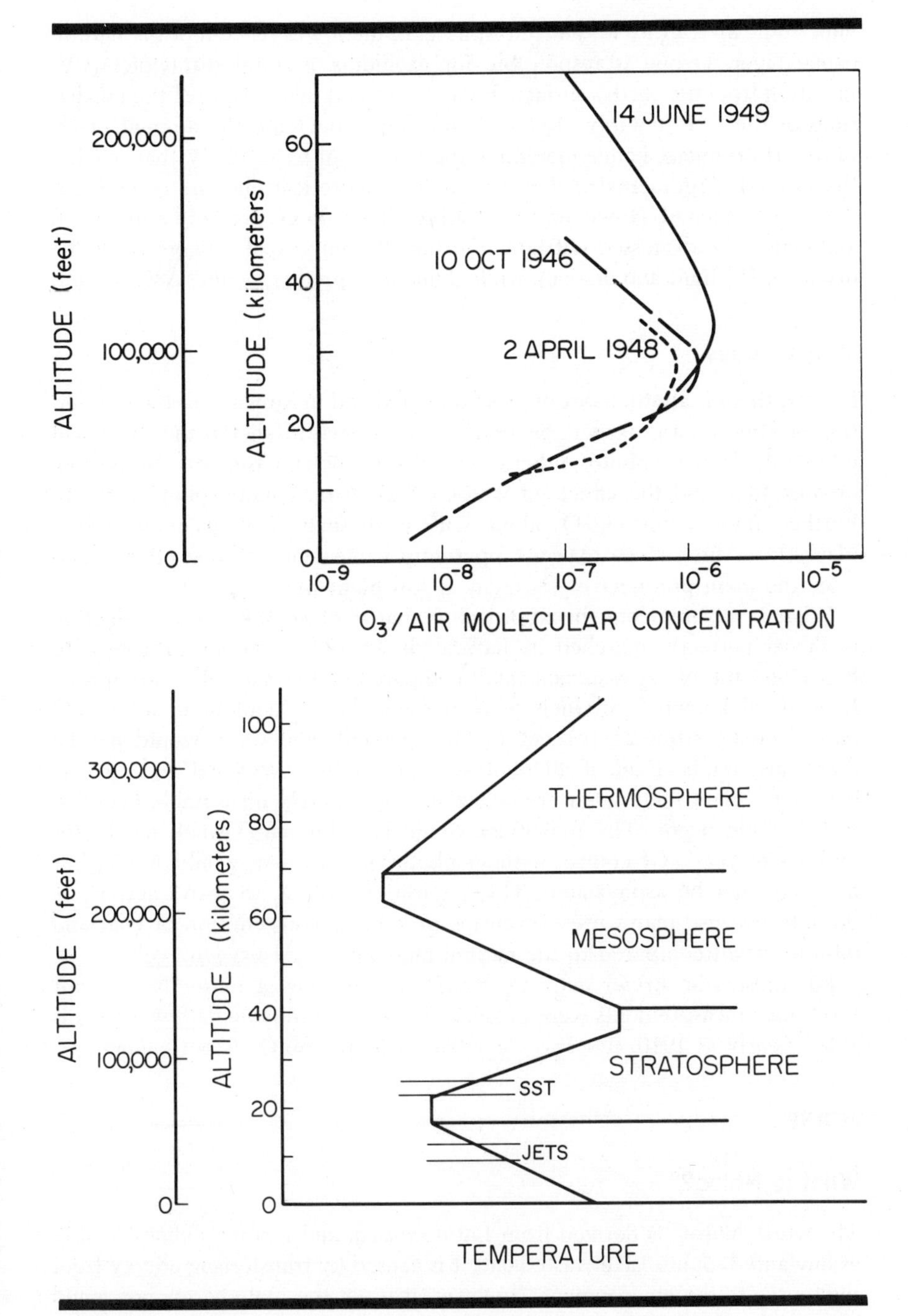

Figure 6–8. Variation of Temperature and Ozone Concentration With Altitude

Adapted from THE UPPER ATMOSPHERE STUDIED BY ROCKETS AND SATELLITES, by H. E. Newell, in PHYSICS OF THE UPPER ATMOSPHERE, ed. by J. A. Radcliffe, Academic Press, 1960.

Supersonic aircraft fly in the stratosphere, in the lower reaches of the natural ozone layer. Ozone is responsible for excluding harmful ultraviolet (UV) radiation from the earth's surface. It can be argued that certain of the emanations of the SST (notably H_2O and NO) might decrease the amount of O_3 in the stratosphere, hence increasing the amount of harmful UV that reaches the ground. Arguments to date are highly theoretical, and no conclusions have been reached. However, the possible effects on O_3 are very real indeed and cannot be dismissed outright. Further, the impact of a decrease of the ozone on UV light and on man, while difficult to predict, could be significant.

Oxygen Loss

Finally, there is another sort of problem embodied in forecasts of the diminution of atmospheric oxygen. Several scientists have predicted that man will ultimately decrease photosynthetic activity on a global basis by deforestation, urbanization, and the effects of water pollution on aquatic photosynthesis. Further, man is burning O_2 along with fossil fuels. This argument seems plausible at first, since oxygen apparently came from photosynthesis, and since the main producer of O_2 today is still plant life.

What is missing from this approach is that photosynthetic O_2 production is almost perfectly matched by natural decay and metabolic processes, so that man's use of O_2 is indeed small compared to the natural consumption. Even if all known fossil fuels were burned, the O_2 content of air would only decrease from 21 percent to 20.8 percent, and these would not be disastrous results. Even if all photosynthetic activity stopped and all carbonaceous plant materials were oxidized, only a barely measurable decrease in O_2 would occur. The remainder would last (for man's small needs) for millions of years. Of course, without photosynthesis man would starve, but he would not be asphyxiated. This insensitivity of O_2 to man's activity is possible because man's uses—even his prodigious consumption of coal and oil—are small compared to the natural amounts in the oxygen cycle.[5]

For those who prefer data, O_2 measurements show a perfectly constant level since measurements were possible. Since fairly reliable data were available as early as 1910, it seems clear that atmospheric O_2 is not changing.[6]

NOISE

What is Noise?

The word "noise" is derived from Latin *nausea*, and is often defined simply as unwanted sound. Like waste heat, it is caused by transferring energy from the source to the environment. However, it is not thermal energy, but sound waves. The amount of energy involved is very small, so that it is an unfortunate but inconsequential loss in terms of the efficiency of the source of noise. The human ear is particularly sensitive to low energy sound waves, and hence a variety of problems arise from noise, ranging from simple annoyance to loss of sleep, deep psychological disturbances, hearing impediments, and safety problems. The brief discussion that follows is intended to introduce the basic problem of noise; the reader is referred to other works for in-depth study. *Noise Pollution*, by C. R. Bragdon, 1970, and *Noise and Man*, by Burns, 1968, are particularly useful introductions.

Sound, including noise, consists of small pressure waves in the transmitting fluid—usually air. The pressure of sound is normally measured on a scale of decibels (dB) which is a non-linear, logarithmic scale. Hence, doubling the sound pressure increases the dB rating by 20 times $\log_{10}$ of 2, or 6 dB. The energy or power involved in noise varies in yet a different way, namely as the square of pressure ratio; the dB level as ten times $\log_{10}$ of the power or energy ratio. Hence, if a jet airplane in a given situation produces 120 dB, two jets will double the emitted energy or power, and the noise level will be only 123 dB. This odd sort of scale is useful to engineers but is very hard for the public to understand, especially since a doubling of a loud noise like a jet aircraft produces a small increase in dB. Table 6–2 includes some examples of energy levels, dB, and typical sources.

Still, decibel levels alone do not suffice to describe noise, since the effect or perception of noise is dependent on the frequency—i.e., whether it is a high or low pitch—and the mixture of frequencies. Measurements of noise, therefore, must include consideration of the spectrum of sound waves in noise, just as we consider the spectrum of light waves in light.

Finally, the time dependence or duration of sound is an important variable. A rifle shot is very short compared to the takeoff of a jet aircraft, and the differences are important to the effects.

Sources of Noise

It hardly seems necessary to describe sources of noise—they are obvious. The kitchen fan, air conditioner, vacuum cleaner, food mixer or blender, and myriad other gadgets produce noise in the home. Industrial noises include metal forming trip-hammers, pile divers, and aircraft wind tunnels. Perhaps the most common sources—and the most continual and unavoidable—are the several forms of transportation in our society. Automobiles, motorcycles, trucks (especially large diesel trucks), trains, and aircraft all contribute to a continual din in cities. In some cities like New York and Chicago, elevated trains running on metal wheels and tracks are a unique and obvious source.

Looking more closely at these transportation sources, we can see that the noise emanates from several sites: exhaust, engine, gear boxes, tires, and in some locations air turbulence caused by the movement of cars at high speeds. Depending on the owners' driving habits and penchants for such noisy gadgets as straight-through exhaust systems, the noise level emitted from cars may be much higher than from other otherwise identical machines. Thus, unlike the case with most pollution sources, it is very hard to state objectively

TABLE 6–2

Noise Levels Measured in Sound Energy

RELATIVE CHANGE IN SOUND ENERGY	DECIBELS	NOISE SOURCE
1	0	Threshold of Hearing
1,000	30	Whispering
1,000,000	60	Conversation
100,000,000	80	Food Blender
10,000,000,000	100	Heavy Traffic
1,000,000,000,000	120	Jet Aircraft

From NOISE POLLUTION, by C. R. Bragdon, University of Pennsylvania Press, 1970. Reprinted by permission of the publisher.

an emission budget for noise. Motorcycles are often very noisy, and it is perhaps in cases such as this that the human being becomes identifiable as the source.

In terms of public concern, aircraft noises are the most important problem. About 6 million people live or work within the noisy confines of the approach and take-off patterns of our large and growing airports. The noise from airplanes (now largely jets) is primarily due to the mixing of high velocity exhaust with the surrounding air, although some of the characteristic whine of approaching planes is due to intake compressors. Newer turbofan engines are quieter than straight jets, and the new generation engines on the Boeing 747, Douglas DC-10, and Lockheed L-1011 reduce noise by 10 to 15 dB, which is very substantial. Remember that 10 dB amounts to a tenfold reduction in sound energy. Increases in air traffic without a corresponding increase in the number of airports accounts for much of the current problem. This factor, coupled with increased urban sprawl in the vicinity of airports, has exposed large numbers of people to the noise of airplanes.

Even though aircraft noises bring the largest number of (and most vigorous) complaints, people are much more uniformly exposed to surface traffic noises. Although automobiles are typically 8 to 10 dB quieter than gasoline powered trucks and 12 to 18 dB quieter than diesel trucks, the large number of cars makes them an important noise source, especially in quieter areas of cities where trucks do not travel. At freeway speeds, the noise of automobiles is mainly due to tires, while at city speeds, the noises come from engines, exhaust systems, and the rattling of loose parts.

Ranking of Sources and Measurement Capability

The complexity and mobility of noise sources, the variability of sound transmission, and the differences caused by operation in varying terrain make it impossible to provide a hierarchical list of noise sources. It is possible with current technology to measure the relevant parameters, and efforts are now underway in a few cities to do so. However, it is necessary to measure more than one parameter: dB, frequency spectrum, and time dependence as a function of location. Expert acousticians then interpret the data both for possible legal or other control action. As a result, noise pollution remains as perhaps the most unknown and unstudied pollution problem.

Effects of Noise on People

The effects of noise vary from mild annoyance to irreversible changes and hearing loss. In addition, noise that interferes with communication often causes accidents, for instance at railroad crossings. Thus there are three basic problem areas concerning noise: psychological, physiological, and interference with communication.

The annoyance of noise is the most general problem, and includes all people in all living and working situations, 24 hours a day, 7 days a week. There is little refuge from noise even in remote mountain areas where motorized vehicles intrude. Decreased comfort, increased irritability, and decreasing work performance are real effects, but difficult to measure. Little is known about the continual din in our environment, but it is obviously a serious problem. The annoyance of noise requires detailed study by psychologists and others to determine what acoustical environment we humans really require.

Besides the psychological and annoyance aspects, there are real physiological effects that can be measured. P. Foster in his *Introduction to Environmental Science* summarized the studies of otologist Samuel Rosen regarding background levels of noise in the United States:

> We seem to have been spared the insult of repeated exposure to supersonic booms, but we have yet to learn very much about the effects of the rather intense levels of ambient background noise that so many of us have become accustomed to in our usual surroundings. A study by a physician makes one wonder. Otologist Samuel Rosen found that members of the Maban Tribes of Sudan, who live in primitive villages where they encounter very little noise, have better hearing than Americans of similar age. In fact, he found that Maban villagers 70 years old have hearing acuity similar to that of 20-year-olds in the United States. Does ambient noise level explain the difference?

Unlike many less severe physiological responses to stress, such as the toughening of callouses or muscles, the human hearing apparatus can be permanently harmed since the long-term impact of noise is cumulative damage. While some sorts of noise result in short-term loss of hearing, continual exposure results in permanent effects. Workers in noisy industries usually have some form of protection, such as ear plugs or soundproof operators' booths, but the public is not usually so protected.

Hearing loss affects a large number of people in the United States. Perhaps 7 million people have hearing loss in one or both ears (Bragdon, 1970). An unknown percentage of these probably lost their hearing due to noise. Many of the 8 to 10 million people who need hearing aids are probably the victims of occupational noise.

Other physiological effects can be attributed to noise, although quantitative data are not at hand. For instance, the stress created by noise results in peripheral vasoconstriction with attendant effects on the heart. Decreases in the oxygen supply to the brain have been found due to high noise levels. Interference with digestive processes has also been noted when a sudden unexpected noise occurs. Still further physiological and psychological effects come from sleep interference.

Perhaps the most significant health problem caused by noise is interference with communication. As Bragdon explains:

> Noise that interferes with communication can be hazardous, particularly when a message intended to alert a person to danger is masked by noise. Two people were killed and several injured the day Senator Robert F. Kennedy's southbound funeral train passed through Elizabeth, New Jersey. They were unable to hear the warning horn of a train approaching from the opposite direction, and therefore did not get out of its way. It was later determined that secret service and news media helicopters had completely obliterated the signal warning of this train's approach.

The warning clang of a railroad crossing, the siren of an emergency vehicle, the call of another person, and an emergency radio message, are all examples of situations where noise may have lethal effects when such signals are masked by other sounds.

In addition to effects on people, extremely loud noises like sonic booms can damage buildings. Windows are especially susceptible to the pressure of sonic booms, as the U.S. Air Force has learned in its flight tests of supersonic aircraft. Damage to structures is less well documented but appears to be real. Even geographical features have been affected by boom-introduced rockslides.

Control of Noise

Noise control, like control of other pollutants, is often best accomplished at the source, although soundproofing is frequently used. Due to the great variety of sources, the technologies are extremely varied. They range from improved mufflers and tires for cars to vibration-isolating dampers for machinery and lead baffles to absorb sound.

The relocating of highways and the finding of alternate sources of transport are needed to suppress the din of automobile and truck traffic. Aircraft noise management has already resulted in alteration of flight paths in urban areas, sometimes despite claims of decreased air safety. Curfews on jet flights at night are also used. Zoning for noisy industries is a solution for new construction, but does not really help existing problems.

In the future, it seems evident that much more work on the problem of noise will be needed, both in legislation and technology. Currently, however, it seems that the racket in our environment will continue for a long time to come, and that noise will probably remain as an important but poorly understood problem.

SUMMARY

Water-borne diseases have reached epidemic proportions in various parts of the world during man's history, but are now largely controlled by treating water supplies. Disease-causing organisms can also be effectively transmitted in the air, but the importance of this pathway to man is relatively unknown. Although they have not been epidemiologically identified as an important pathway, solid waste disposal areas are a potential food reservoir for well known human disease carriers such as rats and flies. Treatment of putrescible solid waste for the control of disease is therefore a highly desirable practice.

Man's contact with chemical toxicants can come from food, water, or air. The effect on humans is a function of total exposure and time, and not merely the concentration in any one food, water, or air sample. If the ingestion rate is significantly greater than the excretion rate, a harmful or lethal level may accumulate. Therefore, the Food and Drug Administration and the Environmental Protection Agency have imposed limits on concentrations of materials in air, water, and food. Pesticides, some heavy metals, particularly mercury, and some radionuclides are examples of toxicants which organisms tend to accumulate. Although many toxicants remain unidentified, intake of organo-chlorine pesticides with food is decreasing, and radionuclide contamination from fallout is presumably less than safe limits, although in some instances food has been confiscated because levels of these toxicants were above FDA limits. "Safe" limits are only estimates, and are contested, reevaluated, and changed as the information base grows.

Of most immediate concern to man's health are toxic chemicals in urban atmospheres. In addition to the direct effects of smog, leading in some cases to death, concentrations of carbon monoxide and sulfur oxides that presently exist in U.S. cities are known to be injurious to health. Sulfur compounds, together with aerosol particles and ozone, irritate the bronchial tract and have the greatest effect on people suffering from bronchial disease, by further restricting breathing, which can lead to emphysema. Carbon monoxide affects reaction time and decreases the oxygen-carrying capacity of the blood. Man is probably exposed to unhealthy levels of other atmospheric toxicants as well, although they are less easily measured.

Man's activities, particularly his intensive use of the combustion process, have had measurable effects on the atmosphere. Most certainly the CO_2 content of the atmosphere has increased, precipitation quantity and quality have been altered, and the air over urban areas is warmer and dustier than over rural areas. The long term significance of these environmental changes is speculative at this time.

References—Chapter 6

1. Dubos, Rene.
Man, Medicine, and Environment. New York: Mentor, 1969.

2. Frank, Robert.
Personal communication and lecture, School of Public Health, University of Washington.

3. Hobbs, P. V., Radke, L. F., and Shumway, S.
"Cloud Condensation Nuclei for Industrial Sources and Their Apparent Influence on Precipitation in Washington State," *Journal of Atmospheric Sciences*, Vol. 27, No. 1, January 1970.

4. Ellis, H. T. and Pueschel, R. F.
"Solar Radiation: Absence of Air Pollution Trends at Mauna Loa," *Science*, Vol. 172, No. 3985, May 21, 1971, p. 845.

5. Broecker, W. S.
"Man's Oxygen Reserves," *Science*, Vol. 168, No. 3939, June 26, 1970.

6. Machta, L., and Hughes, E.
"Atmospheric Oxygen in 1967 to 1970," *Science*, Vol. 168, No. 3939, June 26, 1970.

Suggested Readings

Bragdon, C. R.
Noise Pollution. Philadelphia: University of Pennsylvania Press, 1970.

Burns, W.
Noise and Man. Philadelphia: J. B. Lippincott, 1968.

Foster, P.
Introduction to Environmental Science. Homewood, Illinois: Richard D. Irwin, Inc., 1972.

7 Control and Sensing of Environmental Quality

Every year billions of tons of waste are discharged into the air, water, and land of the United States. Previous chapters have described some of the sources of these wastes and their impact. Automobiles discharge to the atmosphere almost 100 million tons of fumes and gases annually; they become solid waste disposal problems when they are discarded. Factories and power plants add their share of smoke, fumes, and particulates. Each year they generate millions of tons of solid wastes, pour over 30 trillion gallons (92,000 acre-feet) of waste waters to rivers, lakes, and oceans and dissipate waste heat in the environment. Each person generates 100 gallons of waste water daily and 5 pounds of solid wastes; he pollutes the air from his home, heating, and car. Residues from harvesting operations such as mining, lumbering, and animal manure total over 2 billion tons of solid waste annually, and the fertilizers and pesticides used in crop production drain into the waterways. The mounting waste products of our nation must be controlled. Can and should they be stopped? What will be the cost of the control? Who will pay for this and who will benefit? These are the questions of concern in the next two chapters.

This chapter introduces the concept of control and the basic elements of an environmental control system—the sensing and regulating mechanisms.

Before control can be effected, environmental quality must be defined and methods must be developed to measure the defined parameters. It is insufficient to say that control is required to provide *clean* air or *clean* water. How clean is *clean*? If clean is defined as pristine environmental quality, how should this be determined? Before man, there was no record of fluctuations of natural pollutants from forest fires, volcanos, natural erosion, and natural biochemical events that released materials to the land, air, and water. Obviously, pristine does not mean distilled water or air composed of only oxygen and nitrogen. The problem is to define what things are to be measured in the environment and then to determine how, where, and when to measure them.

The next step in achieving control is to define and measure the materials that are discharged into the environment and define their relationships with the accepted measures of environmental quality. Any discrepancy between the actual and the acceptable levels requires regulation.

The following sections will discuss the concepts of control and sensing technology.

CONCEPTS OF CONTROL

A large body of knowledge has been developed in an attempt to understand how systems can be regulated to provide desired results. Typical applications of control theory include the development of control systems to simplify or replace manual control. For example, the automatic thermostat in a home heating system is a control system developed to eliminate the need for a person to regulate a heater to maintain a constant temperature in the home.

To illustrate the basic elements of a simple control system, consider the steering system for a car shown in Figure 7-1. The driver is the controller and is constantly comparing the direction of the car with the direction he would like to move. Any time the actual direction deviates from the desired direction, the driver will attempt to correct this by turning the steering wheel.

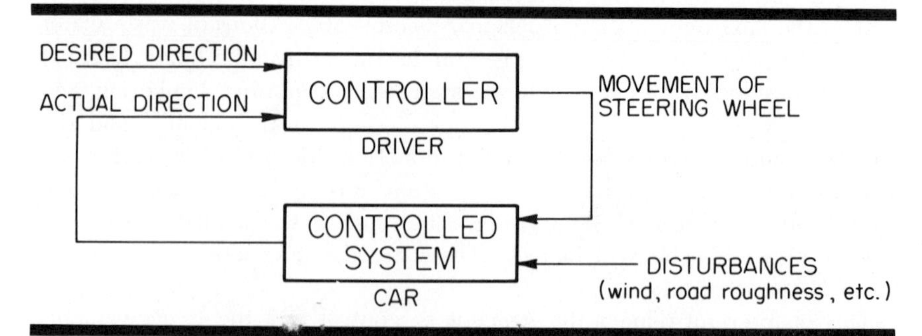

Figure 7-1. Automobile Control Systems

In this example, the driver not only senses the direction of movement, but also determines the desired direction, and makes necessary corrections to effect the desired results. In more complex systems, different persons, institutions, or systems may be responsible for each of these control elements. The problems of such an arrangement can be demonstrated by assigning the control of the car steering to a husband who is driving blindfolded, a nearsighted mother-in-law who is directing from the back seat, and a wife who is knitting and looking to see where the car is going. The importance of the wife's role as sensor is obvious, since without information on where the car is heading, the driver can only steer by impact. Without the direction of his mother-in-law, the husband can avoid collisions but will not follow the desired path. This example emphasizes the importance of sensing and direction and presents the new problem of timing. If the wife does not immediately report the direction of the car and the mother-in-law delays in directing a change, it may be too late to avoid a collision. Feedback information must be timely in effective control systems.

In environmental systems, feedback may be delayed, and when this delay occurs additional predictive control measures must be provided, so that the reaction does not come too late. For example, the use of DDT was in full swing before its bad effects were known. If all species were killed before the use of DDT could be stopped—delayed feedback—the feedback system would have been improperly designed and would have failed.

A feedback system can be improved by providing anticipatory control with some device that predicts future conditions and senses current conditions as well. This type of control attempts to regulate input before change occurs. The clock switch on a thermostat is a simple example. This switch indicates that more heat is required at night and turns on the heater before the cool periods of night; thus the room temperature does not drop before heating begins. Such a control system attempts to stabilize the output of the system and avoid temperature fluctuations that occur with the slow responses of less sophisticated control systems. Thus if environmental quality is to be maintained at some uniform level, not only is *environmental quality sensing* necessary but in addition the *capability to predict future changes* is essential.

Control of environmental systems is not as simple as steering a car or warming a room. Hypothetical environmental systems can be formulated as simple control systems, but real systems have unidentified inputs and feedbacks which frustrate attempts to design control systems. If input is not regulated, the behavior of the system will be difficult to predict. In fact, Elton argues that the more diverse and complex an ecological system, the more stable it is,[1] for in a very complex system there are many subcomponents which provide system redundancy at all levels. When one subcomponent is stressed, others will compensate so that the net effect of stress on the entire system is small. Control by diversity and redundancy is an alternative that should not be neglected. Many problems with the environment are related to man's desire to create single component systems. Crops are planted one at a time; urban areas are uniform; waste treatment favors single large facilities; industries are single purpose. This attempt to produce uniformity creates unstable systems which require large amounts of attention and energy to prevent total collapse.

ENVIRONMENTAL CONTROL POINTS

This section reviews what can and should be measured if environmental quality is to be controlled. When environmental quality is mentioned, it is usually assumed that the appropriate place to measure quality in the environment is known. There may be a number of equally significant points of measurement. A simple system, shown in Figure 7–2, can be used to illustrate possible points of measurement. There are four components in this system connected by inputs and outputs of waste. Removal of component 2 represents the simplest case where wastes from human activities are indiscriminately released to the atmosphere.

Other parameters that can be correlated with environmental quality are waste production, waste discharge, and the social impact of the whole activity. As shown in Figure 7–2, each of these parameters is directly related. To provide effective control, we would have to enjoy a thorough understanding of each parameter. Unfortunately our understanding is severely limited. We often know a great deal about how to produce a material such as cellulose, aluminum, or copper; we know less about the chemical composition of wastes emanating from plants producing these materials, even less about the effects of waste discharge in the receiving environment, and almost nothing about the social impact of the whole process. It must be emphasized therefore that decisions regarding where to control wastes—in the plant, the receiving environment, or the market place—must be made without sufficient scientific understanding.

For example, if domestic wastes were to be controlled, the waste system could be specified first by monitoring the source, then treating the discharge of the waste, and finally, treating the environment receiving the waste, or the aesthetic and social impact of the environmental changes associated with the waste. The control of the system could be based on any parameters shown in Figure 7–2. In this example, there are as many control points as sensing points. When providing control for such a system it is not sufficient to provide one control and one measurement. If waste discharges were selected as the point of observation and control in a total system, and if the control were defined in terms of the concentration of a waste characteristic such as the BOD_5 load, it is quite possible that little remedial control of environmental quality might occur.

If the control statement is defined as "no waste discharge can contain more than 100 mg/l BOD," then a discharger with a waste stream of 500

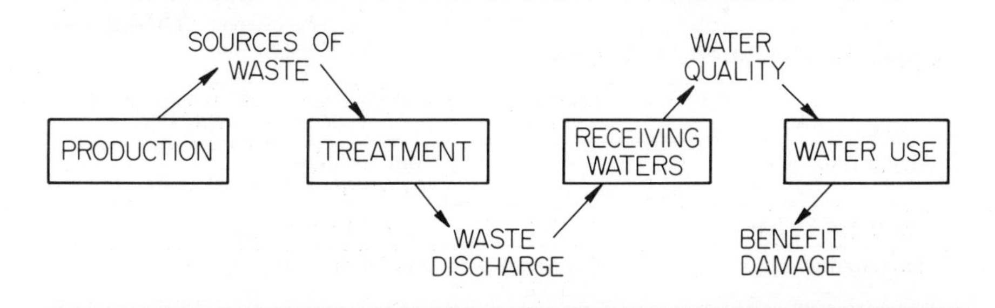

Figure 7–2. Environmental Control Points

mg/l BOD has at least two alternatives which would satisfy this control criterion: he could provide waste treatment that would reduce the BOD_5, or he could withdraw and use water with little or no BOD_5 content to dilute his waste stream to a concentration of BOD_5 less than, or equal to, 100 mg/l. While the latter action satisfies the control criteria, it does nothing to reduce waste discharge to the environment or improve environmental quality. Another example of insufficient control is to define control as "a waste water must have 75 percent of the BOD removed." This type of control will improve environmental quality only if waste production does not increase and negate the impact of treatment. When developing control strategies it is important to insure that the parameters used to assess the effectiveness of the control action also reflect the overall goals of control.

SENSING

Measurement of Air Pollutants

There is a wide variety of compounds in air that need to be measured if proper surveillance of air quality is to be made. Unfortunately, it is usually impossible—mainly for economic reasons—to measure as many things as frequently and as accurately as is necessary, so the job is only partially done. The result is much guess work and a great deal of uncertainty about the effectiveness of control strategies.

Tables 7–1 and 7–2 show the air pollutants most commonly measured, along with a variety of other substances that should be measured but normally are not. Usually, an urban monitoring operation includes only 3 or 4 gases, a sampler for particulate matter, and a filter paper sample to judge the soiling capacity of particles in air.

In some locations, one or another pollutant may be absent or unimportant, which may justify its not being measured.

Air monitoring is usually done from a fixed station, sometimes a parked trailer or van. Air pollution is rarely mapped with mobile ground vehicles

TABLE 7–1

Frequently Monitored Air Pollutants

NAME	FORMULA (if any)	NOTABLE EFFECTS IN EXCESSIVE CONCENTRATION
Sulfur Dioxide	SO_2	Plant Damage Respiratory Tract Irritation
Hydrogen Sulfide	H_2S	Odor of Rotten Eggs
Mercaptans	CH_3SH & Other	Odor of Pulpmills
Nitric Oxide	NO	Photochemical Reactions in Los Angeles
Nitrogen Dioxide	NO_2	Photochemical Reactions in Los Angeles
Ozone	O_3	Los Angeles, Respiratory Tract Irritation
Carbon Monoxide	CO	Asphyxiant at High Levels
Hydrocarbons		Reaction with O_3, Plant Damage
"Total" Particulate Matter		Soiling, Visibility Loss. Possible Health Effects.
a) Mass b) Filter Paper Stain c) Some Elements and Ions		

TABLE 7–2

Less Frequently Measured Air Pollutants

NAME	FORMULA (if any)	EFFECTS
Carbon Dioxide	CO_2	Gross Index of Human Activity, Climate Effects
"Total" Particulate Matter *Molecular* Character		
a) Inorganic b) Organic		a) Toxicity, Plant Damage b) Carcinogenicity Plant Damage Eye Irritation

or aircraft. The location of the fixed site or parked vehicle is obviously crucial to the results obtained. Locations are chosen to be representative of air quality zones. The zone of highest pollutant concentration in a region has the highest priority for monitoring. Areas with high population density are also selected for measurements. Other stations are supposed to be located on the periphery of a region to assess the quality of air there. Areas of projected growth are sometimes measured to determine the impact of increased human activity. Finally, some uniform sort of coverage of a region is desirable. Even though these are the Environmental Protection Agency's official guidelines for locating monitoring sites, there is good reason to believe that location of air pollution sensors will remain a major problem for some time to come. There are several reasons:

1. Most areas cannot afford this extensive coverage and try to make do with an insufficient number of sites based on EPA guidelines.
2. Some cities may not have representative locations because of high population density and heterogeneity—Manhattan and downtown Chicago are examples.
3. Unlike water in a well mixed river, air composition varies in three dimensions spatially as well as in time. Complex air currents transport pollutants to receptors in all three dimensions. Only when some systematic knowledge of air movement is obtained can proper control measures be initiated. No city in the United States today has three-dimensional air pollution mapping, although California is beginning to work in that direction.

Figure 7–3 shows the spatial distribution of one pollutant—SO_2—in Manhattan that typifies the problem of siting monitoring stations. Figure 7–4 shows a contour map of visibility degradation in Tacoma, Washington, downwind from a large industrial point source. Because these patterns change with wind conditions and height, it is obvious that care must be taken when interpreting air quality records from one site.

Measurements of pollutants sometimes are made continuously, although all too often they are done only 5 days a week and sometimes only 8 hours

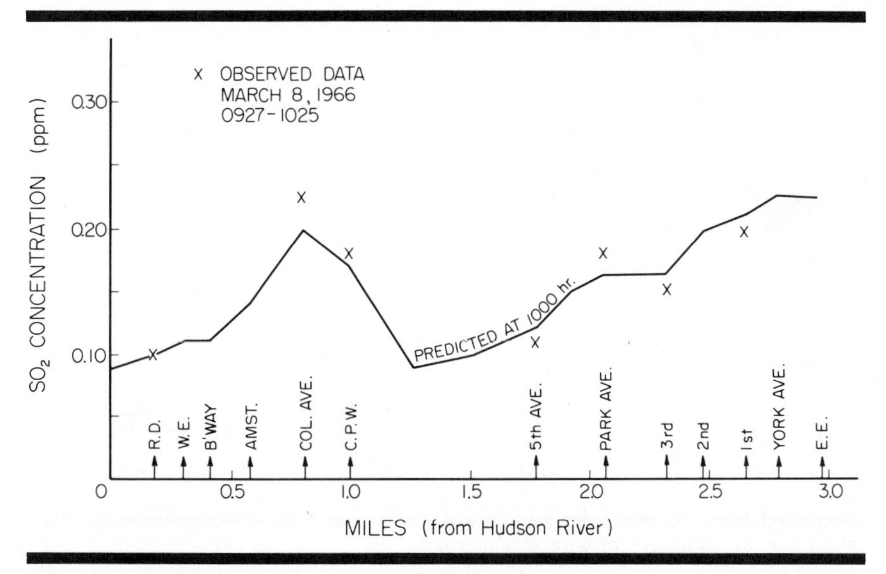

Figure 7–3. Spatial Distribution of SO_2 over Manhattan

From L. J. Shieh, Ph.D. Dissertation, Fig. 38. Department of Meteorology and Oceanography, New York University, 1969.

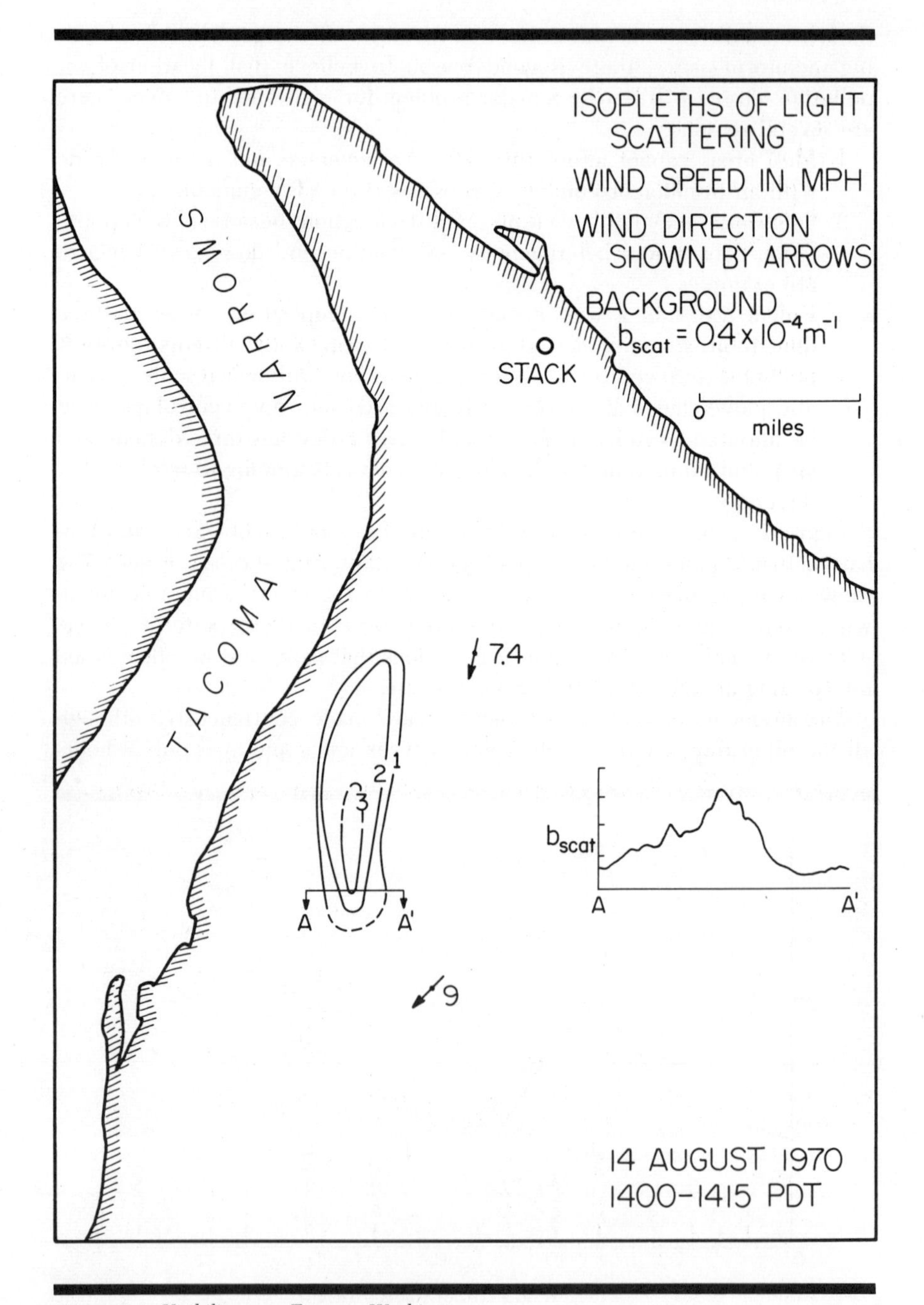

Figure 7–4. Visibility over Tacoma, Washington

From D. J. Lutrick, Thesis for Master of Science in Engineering, Department of Civil Engineering, University of Washington, 1971.

a day. Sampling particulate matter is usually done on a 24- or 48-hour basis. Only continuous monitoring can provide reliable statistics on the behavior of a pollutant at a given site. Usually several months or years of data are needed to provide the necessary information on the site. Often a decade is required to establish trends. Unfortunately, not all air pollution officials and politicians understand this fact, and they attempt to establish trends based on insufficient data. This tendency may be especially pronounced when air quality improvement is politically helpful. Figure 7–5 is a sample of time dependent data for several pollutants, and clearly shows the need for continuous rather than instantaneous measurements.

It is possible to get crude, limited air quality data with instruments costing a few hundred dollars. Usually, such instruments also are slow and expensive in terms of man-hours, so that it requires more time to establish the characteristics of the site and even longer to establish trends. Faster-responding, more accurate, automatic devices are expensive—usually several thousand dollars each—with a separate device required for each pollutant. Nonetheless, many large cities have decided to take this course even though the cost

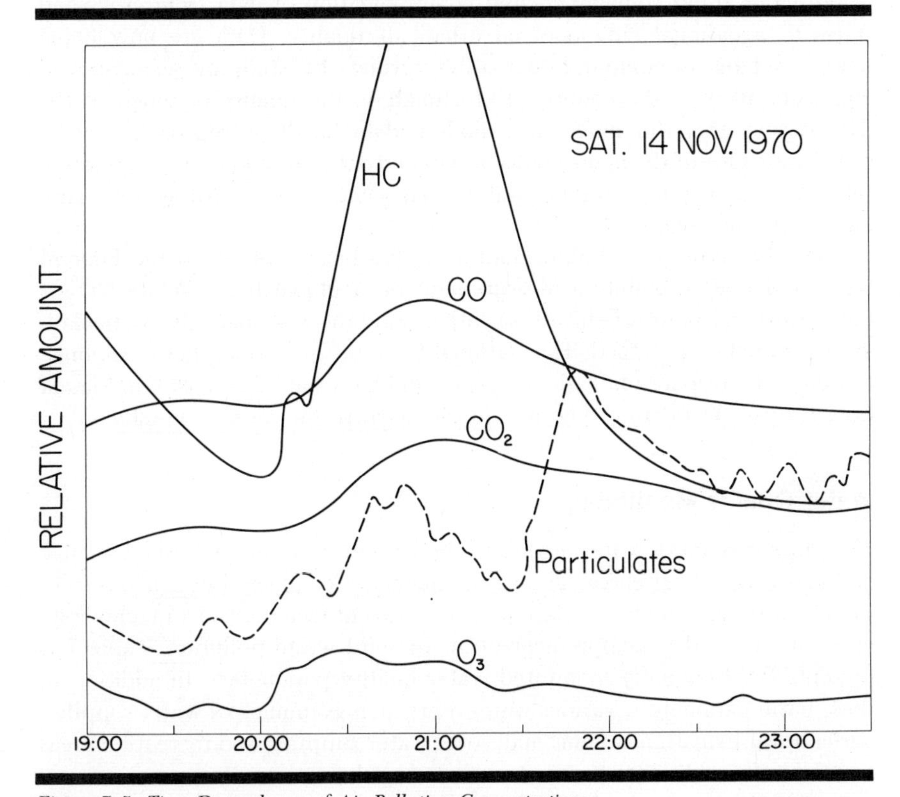

Figure 7–5. Time Dependency of Air Pollution Concentrations

of equipping a single station for, say 6 gaseous pollutants, is around $30,000. Since a city such as Los Angeles requires a large number of sites, the cost of instruments alone is high, perhaps approaching a million dollars. Few cities have such resources available to them and have to build up their stockpile of instruments over a long time.

One of the major problems that exists now, and probably will exist for some time to come, is the cost of renovating outmoded instruments. The high capital cost precludes rapid conversion to automatic devices, so many monitoring organizations are trapped in a worsening situation with high personnel costs needed to operate manual instruments.

Air pollutant measurements are made by government agencies at all levels and by some private industries as well. Many city and state programs are assisted financially by the federal government. There has been a shift in responsibility for air monitoring during the decade 1960 to 1970 from public health agencies to environmental control agencies. There has also been a gradual shift toward more Federal support—and therefore Federal control—of monitoring programs.

With this trend comes increased communication of data from cities and states to a national data pool on urban air quality. Data are now being made available in computer-compatible formats for study by scientists and engineers all over the country. Unfortunately, the quality of much of the data is such that this move to a modern data handling approach may be premature. One of the main results of detailed data study so far is a suggestion that the old, manual methods—and even some of the earliest automatic ones—are not reliable.

One other type of pollutant monitoring has begun, wholly at the Federal level, and that is pollution measurement on a global basis. While CO_2 is the main focal point of this effort, other substances—especially particulate matter—are being studied. The National Oceanic and Atmospheric Administration is the responsible Federal agency and has established a site on Mauna Loa, Hawaii. Data from this site are discussed in the section on effects.

Water Quality Monitoring

The measurement of water quality in the laboratory is an established science and many analytical chemical measurements are conducted in aqueous solution. Determination of materials in water does not face the lack of technology encountered in the measurement of air or solid waste pollution. Table 7–3 presents the frequently monitored water quality parameters. In addition to these basic parameters, various water users such as municipal water supplies, agricultural irrigation systems, industrial water supplies, and recreation areas have specific additional parameters that must be measured.

One of the major problems of water quality monitoring is how to sample a body of water in terms of frequency and place. Historically, water quality

TABLE 7–3

Typical Water Quality Standards of Selected States

	BACTERIA/100ml	ODOR	DISSOLVED SOLIDS, mg/l	RADIOACTIVITY, mmc/l	SPECIFIC CHEMICALS	pH	TEMP., °F	DISSOLVED OXYGEN, mg/l
Kentucky	5,000	3	500	10^3	X	5-9	93	5
Illinois	5,000	—	500	10^3	X	6-9	93	5
Indiana	5,000	—	500	10^3	X	5-9	93	5
Pennsylvania	1,000	24	500	—	X	6-8.5	87	5
West Virginia	1,000	24	500	10^3		6.5-8.5	93	3
Tennessee	10,000	—	500	—		6.5-8.5	93	5
Alabama	1,000	—	—	—		6-8.6	90	4
Mississippi	5,000	24	500	10^3		6-8.5	93	4
New York	—	—	—	—		6.5-8.5	—	5
Maryland	5,000	—	—	—		6-8.5	72	6
Florida	1,000	—	—	—		6-8.5	—	4
Washington	1,000	—	—	—		7.5-8.5	65	8

From Design of Water Quality Surveillance Systems
Federal Water Quality Administration Report 16090 DBJ 68/70
United States Department of the Interior, August, 1970.

data have been obtained only at a few locations along a river on a monthly, quarterly, or yearly basis. A person would be dispatched to the location and would dip a bottle in the river to obtain a sample. Usually the location selected was a bridge to permit sampling near mid-stream, in other cases samples were "grabbed" from the bank. In no way could such samples be statistically significant or scientifically representative.

The problems of collecting samples for laboratory analysis can be avoided by the use of remote monitoring systems. Such systems obtain samples by pumping water from the river to an automatic analyzer located on shore, or automatic detectors can be positioned in the stream to transmit data to a recorder. These systems are expensive to install, operate, and maintain. In addition, such systems are the target of vandalism and subject to a high rate of destruction. In spite of these problems, a continuous remote monitoring system is required to detect the fluctuation of water quality. For example, water temperature can vary more than 10°F in a day in the summer, oxygen content in a stream can fall to levels lethal to aquatic life in a zone a few miles long, or an accidental release of toxic waste can flow by in a few minutes.

A comprehensive water quality monitoring system defined by the federal government has subsystems of data acquisition, analysis, transmission, handling, and use. Such a system employs on-site analysis for a few easily measured parameters and periodic laboratory analysis for detailed water quality studies. Information from both analyses is transmitted to a data bank, reported on standard statistical forms and special reports, and retained in a major data bank. The goal of the Federal Water Pollution Control Administration in 1970 was to collect limited data for a long time period at 400 points in this country and make short term observations at 20,000 points. STORET is a computer-operated storage and retrieval system for water quality data that is operated by the Environmental Protection Agency. Most of the fragmented water quality data available in this country has been placed in this information system.

The ORSANCO surveillance system for the Ohio River represents an advanced monitoring system utilizing electronic monitors that measure several chemical and physical parameters. These data are transmitted in sophisticated real-time data systems and can predict as well as measure water quality. However, the man with a bucket and a thermometer is still a common surveillance system. In the future, remote observation and sample collection employing aircraft will become common practice, and satellite surveillance employing color and infrared imagery will be developed to monitor water quality.

Solid Waste Monitoring

Determining the composition of solid waste is not a common operational activity in solid waste management systems. Many operations borrow laboratory batch sample results reported in the literature as a measure of their own waste composition. In many situations, the quantity of waste is noted by the number of truckfuls, while in others the actual weights of waste are reported. There have been proposals for standard accounting systems for solid waste monitoring but no common criteria currently exist. Monitoring of solid waste is in its infancy and currently is a research curiosity rather than a management activity.

Ecological Methods of Sensing

Water quality management practice to date has required the determination of safe levels of toxicants in waste discharges. When "safe" levels of toxic chemicals are estimated, consideration of additional factors in the environment becomes necessary. For example, existing water quality conditions, such as the amount of calcium and magnesium, can significantly influence the effectiveness of given concentrations of toxic metals. If water hardness is high enough, the toxicity of the metal can be inhibited. On the other hand, if two toxicants occur simultaneously the total effect of these two toxicants can be greater than the sum of their two effects independently. Furthermore, the response of the test organism varies considerably depending upon the species involved, its life stage, and the duration of the test. Other characteristics that are important are temperature, pH, and oxygen content. These latter factors not only affect the activity of the organism and indirectly determine the effect of the toxicant, but they can also control the amount of the metal that is in the solution. Most metals must be in true chemical solution to be toxic. For example, water at high temperature, low pH, and low oxygen content, probably will have more metal in solution than water at low temperature, high pH, and high oxygen content. It is possible for one toxic waste effluent to cause a fish kill in one receiving water and yet produce no effect in another. It would serve little purpose to present the concentrations that are considered lethal to various species because the variations in lethal toxicity levels of any given metal are large.

Toxic wastes can be controlled by predicting potential effects through biological tests known as bioassays. This procedure involves determining the lethal effect of various concentrations of wastes on important organisms that occur in receiving waters. From an experimental relationship as shown in Figure 7–6, a TLm (median tolerance limit) can be determined—that concentration at which 50 percent of the test organisms die in a standard exposure period, usually 4 days. By evaluating the relation of waste concentration and percentage of survival of test organisms, "safe" levels of the waste, or the maximum that would presumably cause zero mortality, can be estimated. An estimate of a safe concentration, together with the total volume and flow of receiving waters, allows calculation of acceptable quantities of waste to be discharged into that receiving water. This is an effective procedure for controlling waste discharges. In many instances, the industry involved is cooperative enough to conduct its own bioassays and apply results to its discharge practices.

As mentioned previously, levels that simply allow survival for a prescribed period of time should not be the ultimate objective of waste management. Bioassay procedures are being revised and improved to include longer periods of exposure, more sensitive measures of effect than mortality (e.g., reproduction and growth), and testing of more sensitive organisms in an attempt to arrive at concentrations that are chronically safe for aquatic life and not merely sufficient to prevent acute mortality. The TLm is conventionally multiplied by a coefficient of 0.1 to determine a "safe" concentration. Long term studies with several types of toxicants have shown that this is not an unreasonable correction. With other toxic constituents a coefficient of at least 0.01 is required.

To illustrate how a toxicant in effluent might be controlled through bioassay procedures, consider waste entering a receiving water that has a 50,000 gallons per minute flow and a toxicant concentration of 1000 milligrams per liter. Bioassays have shown that the toxic constituent in this waste has

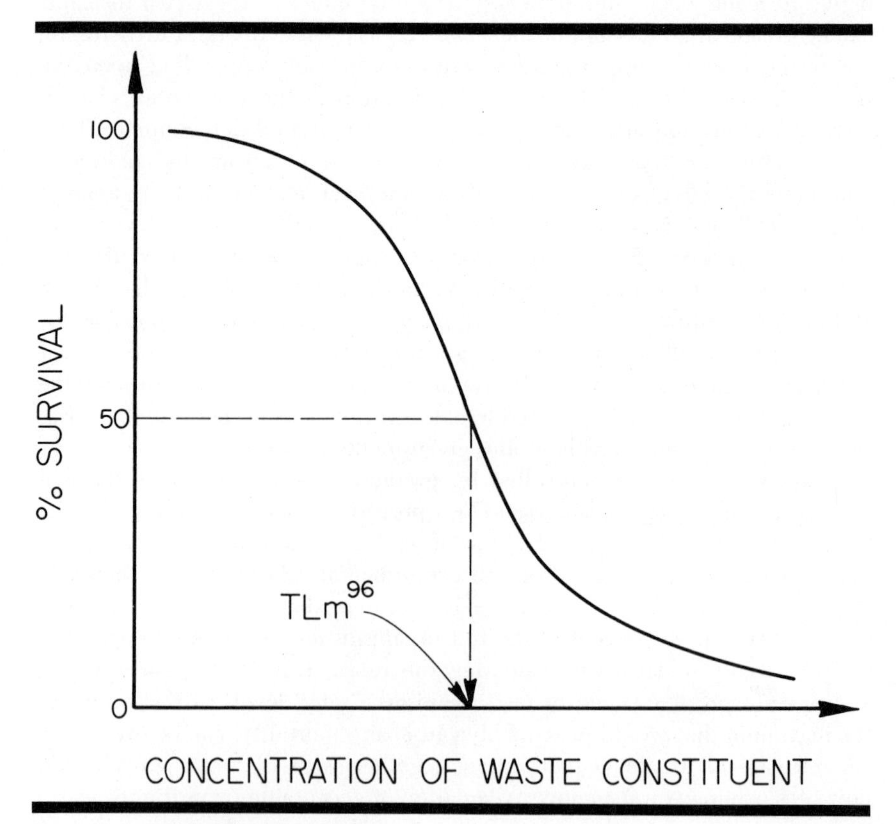

Figure 7–6. Median Tolerance Limit

a TLm of about 1 milligram per liter. If we divide 1 milligram per liter threshold by 10 to allow a safety factor for possible effects that could occur from continuous exposure, we arrive at a "safe" level of 0.1 milligram per liter. The allowable discharge rate of that effluent would be 5 gallons per minute and, assuming complete mixing, this would produce a concentration of 0.1 milligram per liter of the toxic constituent in the receiving water.

The bioassay method for regulating waste effluent discharges is effective for wastes with recognizable components but it makes its most important contribution in controlling complex wastes that have many factors which cannot be easily separated and whose constituents actually have synergistic effects on toxicity. Pulp mill wastes are examples of such complex wastes and mill discharges are regularly controlled through bioassay procedures. Bioassay test organisms usually are those that have economic importance, such as sport or commercial fish. In Puget Sound, shellfish (oysters) are important economic aquatic species, consequently, extensive bioassay procedures using larval oysters have been developed and the method shows great promise.[2]

Generally, water quality standards do not apply to effluents; that is, they are set so that the concentration level is achieved after mixing of the effluents with the receiving water. Most standards allow a portion of a river, for example, as a mixing zone for dilution of the waste before the standard is applied. The length, breadth and depth of the mixing zone has often been a point of controversy. Particularly with regard to temperature, regulations stipulate that the mixing zone should not take up more than 66 to 75 percent of the cross-section of the river. The area not affected will allow passage of migratory species and ample downstream passage of food organisms suspended in the water. The total area that is encompassed by the effluent-river mixture is quite important in determining the overall production of aquatic life and may be more important than a "safe" concentration of toxicants.

Although bioassays represent the best available biological tool to predict the effects of toxic wastes on specific organisms in receiving waters, the effects on the structure and productivity of total communities is nearly impossible to predict with any precision. As shown in Figure 7–6 only general trends in the structure of bottom organisms in aquatic environments are predictable after additions of toxic material. Whereas measurements of chemical constituents of wastes in receiving waters are often inadequate for determining the waste effects, because of daily variations, the community of organisms itself is very sensitive to the integrated effects of wastes. If significant changes in the biological community are not apparent, then waste concentrations are low enough not to be detrimental. If, however, a significant change occurs, the relative concentration of waste can be estimated from the degree of change. To correct the situation through waste treatment requires knowledge of the waste constituents causing the damage and how much reduction is necessary. This is very difficult to determine, particularly if more than one waste discharge is involved. Although many dangerous situations have been greatly improved or eliminated by this approach, it should not be relied upon. That is, the environment should not be used as a laboratory to determine "safe" levels of waste discharge. There are two reasons: first, the environmental damage that occurs during the test period should be an unacceptable cost; second, and more important, environmental damage may have proceeded too far by the time it is detected for recovery to be prompt and complete. Thus, a very conservative approach should be taken at the outset of a waste discharge program.

SUMMARY

Control theory provides a basis for examining past failures of environmental quality control efforts and can be used to indicate necessary actions to avoid future failures. Based on this theory, three conditions must be satisfied if control is to be achieved. First, a target condition must be specified, such as some level of environmental quality. Second, a means must be available to sense the actual condition and compare it with the desired target condition. Third, methods must be available to correct the actual condition. *Feedback* is a control process in which deviations from desired levels are sensed and used to automatically adjust a system to some desired target level. Environmental control has been historically characterized by too little feedback. For example, by the time DDT was recognized as a hazard to the environment, normal corrective measures were ineffective and major institutional changes were required to have any impact. This approach generally leads to a series of "stop and go" control policies which can strain the political stability of a nation. An alternative control policy would be to develop prompt sensing of small deviations from desired quality levels so that minor and rapid corrections could be instituted. Another concept of control would involve placing only a small fraction of the total environment under stress and maintaining a diverse and redundant ecosystem which would be more resistant to minor local perturbations. Mass production, economies of scale, and rapid growth are not compatible with such a control concept.

Sensing appears to be the critical element of environmental quality control and has been ignored to date. Not only must sensing of environmental quality be developed, but the sensing of actual waste production and discharge must also be achieved. Although water quality has been observed for many years, much of the data have been obtained at the wrong place or wrong time. Air quality monitoring is in its infancy, and only a few pollutants can be detected. Solid waste data are so meager and fragmented that control is difficult. Faced with this lack of sensing technology, the *bioassay* or observation of environmental conditions lethal to specific organisms may be the most expedient tool at present to establish "safe" levels of environmental quality.

Finally, environmental quality is a difficult goal to define. The conflict between economic growth and environmental quality will continually shift the target condition. Not only is rapid sensing necessary, but a flexible control process is also required to accommodate shifting goals of control.

References—Chapter 7

1. Elton, C. A.
The Ecology of Invasions by Animals and Plants. London: Methuen & Company, 1958.

2. Woelke, C. E.
"Measurement of Water Quality with the Pacific Oyster Embryo Bioassay." Water Quality Criteria, ASTM STP 416, American Society Testing Materials, 1967.

8 Methods and Costs of Waste Management

The American economy is based on planned obsolescence and the rapid conversion of a product to waste. Environmental control cannot merely treat waste from production activities; it must also reduce the volume of products consumed and regulate the ultimate disposal of waste in the environment.

SOURCE CONTROL

Product Modification

Three alternatives can be used to control waste: modification of products, modification of processes and types of raw materials, and elimination of unnecessary products. Product modification is accomplished when products or wastes are identified as serious environmental contaminants. Detergents provide a good example of product modification for environmental quality control. When first introduced, detergents were not biodegradable and persisted for a long time in receiving waters. This resulted in unsightly foam on many waterways, in treatment plants, and in drinking water wells, as shown in Figure 8–1. The development of treatment facilities to remove foam was rejected in favor of modifying detergents so that natural degradation

Figure 8–1. Foam in Waste Treatment Plant

could occur. Conversion involved an annual production of 400 billion pounds of detergent and hundreds of millions of dollars invested in research and new equipment. With the resolution of the foam problem, detergent manufacturers next faced phosphorus pollution. Phosphorus is a primary component in detergents as well as one of the principal limiting nutrients that contribute to the eutrophication of lakes and streams. The current dilemma is whether to treat waste water to remove phosphorus or have detergent manufacturers use a substitute for phosphorus in their product.

The successful product modification of detergents is a matter of history. There are now several other products undergoing study for modification rather than treatment in our waste waters. DDT and BHC (benzene hexochloride) are not easily decomposed. It has been estimated, that nearly two thirds of all DDT ever produced still remains in the environment, assuming that measuring methods measure only DDT. Most of this will eventually reach the sea and be incorporated into and transferred through the ocean food chain. Concern over the long term effects of BHC and DDT has resulted in a noticeable decline in their use. One of the reasons for this decline has been the development of new and better pesticides. A rough estimate of the economics of pesticide modification can be observed from pesticide sales, approximately $500 million compared to $70 million per year for research on the control and effects of pesticides. Actually, the sales of DDT or chlorinated pesticides have declined from about 300 million pounds per year in 1960 to about 250 million pounds per year currently. The decrease in the use of chlorinated pesticides has been brought about by the development of degradable and more specific pesticides. The development and production of degradable, short lived DDT type compounds by modifying compound structure is a promising area of research and could be a boon to pesticide users, manufacturers, and hopefully the environment. Thorough testing and evaluation are needed to assure that new compounds do not have environmental effects as deleterious as those of DDT.

Product modifications remain to be attempted in food preparation, design of products for longer life, and designing of reusable products. Food wastes have been reduced by preparation of foods before sale. Removing husks, greens, and peels from food at the farm allows these materials to be returned to the soil and reduces the solid and liquid waste generated in the home. Peeled potatoes that are cut and ready to cook are another example of product modification. Obviously, if products lasted longer there would be fewer to discard and to produce. Increased product life could reduce waste significantly. Reducing several product components to one increases the possibilities for reuse. All-aluminum cans can be recycled. Simplifying automobiles to eliminate copper and other nonferrous metals from the body would facilitate the reuse of scrap metal. Another example of product modification is the removal of sulfur from fuels to reduce the emission of SO_2 when the fuels are burned.

Process Change

Given the desire to reduce waste but not the ability to modify products, many industries can modify manufacturing process to reduce the amount of contaminant produced. In many cases, this type of control is much more effective and more economical than waste treatment. As an illustration, a pulp mill obtains pulp by removing one-half of the tree and capturing the cellulose. In the simplest process, one-half of the tree is discharged as waste. Because of economic incentives pulp mills may attempt to recover this waste and reuse the chemicals for the pulping process. Technology can reduce the waste from pulp mills to less than 5 percent of the tree, approximately a tenfold decrease in waste.

In many industries, good housekeeping—minimizing spills, sealing leaks, etc.—can significantly decrease the amount of waste discharged to the environment. In many cases, good housekeeping and waste reduction are directly related to increased product recovery and reduced requirements for water intake. Both of these benefits provide economic incentives for waste control by process modification.

The sugar beet industry is a good example of process modification. Sugar beets are washed and transported in a water flume into a process area where they are sliced into small noodle-shaped strips. Sugar is removed from the slices by hot water; the resulting juice is clarified by adding lime and carbon dioxide. After filtration the juice is concentrated into a thick syrup and evaporated in a vacuum to form a suspension of sugar crystals in a heavy syrup. The crystals are separated from the syrup in a centrifuge, then dried and stored. The heavy syrup still contains sugar so it is recycled for further evaporation and crystallization. The final liquid residue from which no more sugar can be economically crystallized is a by-product, molasses. Other by-products can be obtained by using the spent beet pulp for animal feed. Adding molasses to this pulp enhances its food potential. Normally, separate streams of water are used to wash the flume, the washer, the diffuser, the filter, the evaporator, and all the other phases of the process. By process modification, water from a previous phase can be used for the next so that the same water is continuously used and reused, then sugar and other wastes which would be carried away by separate streams can be extracted later in the process. This process modification results in a reduction of the BOD discharged.

The Steffens process was developed to recover an extra 10 percent of sugar per ton of beets by treating the waste molasses. Molasses is first diluted to a concentration of about 7 percent, then pulverized lime is added. Under optimum conditions, this process removes about 85 percent of the sugar from the molasses and increases the amount of product removed from the wastestreams.

More than 50 percent of the BOD wastes can be reduced by process changes in silo drainage and Steffens waste recovery units. By using driers and evaporators, and recovering the drainage liquid from these silos, additional sugar can be recovered and a waste problem is averted. By recirculating the water, thereby allowing additional sugar to be extracted, an additional one third of the waste stream can be recovered rather than discharged to the environment.

There are numerous other examples in the food industry where significant process modifications have drastically reduced or minimized waste discharge to receiving waters. Examples such as improved use and collection of meat packing waste have demonstrated reductions in waste by factors of 2 to 3. As mentioned earlier, shifting pulp mill processes from sulfite to Kraft can reduce waste discharges by a factor of 7. And finally, a very significant reduction in BOD waste was observed when tanning industries converted from chrome to vegetable-base tanning. In this case, a 25-fold reduction in waste was achieved.

Some very significant process modifications can be instituted to reduce air pollution. One instance is the conversion from fossil fuels to nuclear fuels as the energy source for large power plants. This change in process and source of fuel has eliminated one pollution problem, but created another—CO_2 and SO_2 were traded for radioactivity.

A third environmental control measure involving source control is the elimination of a product entirely. New products can be developed which perform the same function as old ones but which are not sources of environmental damage. An example of elimination of product and substitution of a new product was referred to earlier in the case of pesticides. Probably the most desired product elimination would be the internal combustion engine. Alternatives could be the electric car, the steam-powered automobile, or multiple passenger units such as mass transport systems. The electric car not only reduces air pollution but also eliminates the need for the transmission hump, radiator, exhaust pipe, muffler, and fuel plumbing system. The electric car, however, would carry half its weight in batteries, and would cost approximately 25 percent more to purchase. However, it would cost less than half as much to operate as a comparable gasoline-powered car. One of the perplexing questions posed by electric automobiles is that they run on batteries and batteries require power to be recharged. The electric automobile might not really eliminate air pollution because recharging electric batteries requires electric power, and if the electric power were generated by fossil fuel power plants, air pollution would not be eliminated—only transferred from one source to another. The possibility that nuclear power may be available could mitigate the electric car dilemma. There is hope also that chemical fuel cells, that is, batteries which do not require recharging but which generate electricity from chemical reactions, may replace conventional wet cell batteries.

Carrying the product elimination concept one step further, the question is whether the automobile should be eliminated and replaced by rapid mass transit or by systems which deliver products to people, thus reducing the need for personal mobility. In addition, future improvements in communications technology may reduce further the perceived need to travel.

In Los Angeles about 7 million gallons of gasoline are consumed each day. The resulting combustion releases about 30,000 pounds of lead a day. There is major scientific disagreement over how much lead humans can tolerate and what the standard should be, but the concentrations of lead near heavily traveled highways exceed 100 micromicrograms per cubic meter. Lead released by combustion can be reduced by designing treatment units to remove lead from the automobile exhaust, or, more economically, by removing lead from gasoline.

Unfortunately, environmental control specialists have not focused on controlling the production of contaminants. It has been assumed that institutional constraints prevented controlling the sources, therefore, emphasis has been placed on treating waste after production. In several documented cases, source control has proven practical and economical. Source control should be one of the first alternatives considered to improve environmental quality.

METHODS AND COSTS FOR WASTE TREATMENT

The conventional approach to pollution control is to treat the waste-stream to remove objectionable contaminants before discharge. Hopefully, every effort has been made to reduce waste production before resorting to treatment. It makes little sense to provide extensive treatment facilities for waste-streams containing pollutants that can be easily eliminated or reduced. The waste treatment process involves allocating effort between three activities: transporting waste to treatment facilities, treatment, and transporting the residue and the treated waste-stream to points of discharge. This division is strongly dependent on the magnitude of the waste source and the physical state of the waste-stream. For example, solid wastes are collected and transported at great expense to low cost disposal sites; liquid waste may or may not be collected and transported for treatment, but it is usually transported for disposal; gaseous wastes tend not to be collected or transported, rather, each source is treated separately.

There are two alternative ways to examine treatment of waste-streams. One is to examine the streams and then discuss methods for removal of the various contaminants. The other is to examine the materials to be removed independently of the stream. We will discuss both methods, the former for treatment cost comparisons, and the latter to demonstrate physical phenomena and available technology.

Treatment Phenomena

The basic problems in waste treatment are: (1) to separate solids from liquids or gases, (2) to separate solids from solids, or fluids from fluids, (3) to separate one material dissolved in another, and (4) to render toxic or pathogenic agents harmless. Whereas exotic and complex treatment mechanisms can be formulated, simple concepts exist that have not been applied. Some examples of the basic treatment problems are shown in Table 8–1. There are some minor differences in the problems associated with the physical state of the media (gas, liquid, or solid), but the basic mechanisms are similar.

Solids Separation

The separation of solids from a liquid or a gas is usually accomplished by employing screening devices or devices that make use of density differences. Screening as a treatment method is conceptually simple—a barrier restrains the solids and allows the fluid (gas or liquid) to pass with minimum interference. The technical problems lie in devising systems to collect and dispose of the solids and in minimizing the energy required to pass the waste stream through the barrier. Solid wastes can be screened prior to further processing and usually are sorted to recover secondary materials. Waste water screening devices include trash racks that remove car bodies, animal carcasses, trees, etc., coarse screens with one or two inch spacing, finer screens as small

TABLE 8–1

Examples of Basic Treatment Problems

BASIC TREATMENT PROBLEM	AIR QUALITY	WATER QUALITY	SOLID WASTE MANAGEMENT
Separates solids from liquids or gases	Particulate removal from combustion devices, and solids handling processes	Suspended solids from domestic and industrial waste waters, gravel washing, etc.	Problem is inverse to remove gases or liquids from solids
Separate solid from solid or fluid from fluid	Control of CO, NO_x, SO_2, HC from emissions	Removal of oil from water or water from oils	Separation of materials for recycles and treatment
Separate one material dissolved in another	■ Aerosol control ■ Scrubbing of gases	■ Remove dissolved BOD ■ Remove dissolved salts ■ Coagulation of fine solids ■ Adsorption of toxic materials	Recovery of pure metals from alloys, recovery of fibers from paper, etc.
Convert waste to an acceptable form	■ Chemical conversion of odors ■ Combustion of contaminants	■ Biodegradation of waste ■ Chemical oxidation of waste ■ Precipitation of dissolved solids	■ Incineration ■ Composting ■ Wet oxidation ■ Pyrolysis

as 1/32 of an inch, and filters that employ sand, coal, or cloth as filter media. Air filters are similar to household vacuum cleaners, except the filter fabric can be selected to withstand heat and corrosion. A fabric filter can remove 0.5-micron particles almost completely and can remove significant amounts of particles as small as 0.1 microns from gas streams. Although the design of screens is more an art than a science, there is no technological breakthrough required to develop screening processes. Improvements can be made to avoid explosions in gas filters, to reduce pressure fluctuations in gas or water streams passing through screens, to clean screens, and to reduce the cost of screening equipment, but none of these are acceptable excuses to avoid providing solid separation treatment.

The other method of separating solids from fluids involves the use of density differences. Anything heavier than air or water will settle out, and anything lighter will rise. Devices that separate solids from other solids using density differences are called fluidized beds. A tank can be filled with solid waste with air blown in at the bottom. Lighter solids will be blown higher than heavy ones and collection at various levels permits their separation. Naturally when the air is stopped all material heavier than air falls to the bottom. This latter phenomenon is employed in both air and water settling devices. By providing a large volume where flowing fluid can slow down, all the heavy particles will tend to sink. Primary treatment of waste water consists of a large tank for settling of solids. Rectangular tanks are 8 to 12 feet deep and up to 300 feet long. Circular tanks ranging from 35 to 200 feet in diameter also are used. Inlets and outlets are designed to prevent turbulence in the tanks that would hinder settling. A mechanism is provided to collect the settled solids for further treatment. Usually these tanks also have skimming devices to remove lighter, floating materials.

Separating particles from air is not easily accomplished by gravity settling. Only when the particles are large and dense is this method practical. To increase the density differences and increase separation of particulates from air streams, centrifugal force or electrostatic precipitation is used. Air cleaning requires cyclones, centrifuges, and electrostatic units to provide differential forces for removing solids. There is sufficient information available for designing these devices; that is, new scientific knowledge is not required. But, developmental data are not as abundant because of insufficient application.

As simple as solids removal appears, this type of treatment was considered to be a significant environmental protection measure in the 1960s.

Other Separations

Separation of one gas from another or one liquid from another is not as simple as separating distinct phases: gas–solid, liquid–solid, or solid–solid. Separating oil from water or vice versa (a two phase problem) can be accomplished by gravity separation and skimming. The American Petroleum

Institute (API) has produced a manual for oil–water separator design. The concept is simple, but data on performance are sparse.

To separate contaminant gases from air, a mechanism must preferentially select one gas from another, or change the physical or chemical form of the gas to be removed. Much of the effort to solve air pollution by gases such as CO, NO_x, and SO_2 has concentrated on reduced production rather than the development of a removal process, such as those using scrubbers. Scrubbers are devices that remove particles or gases from air by using a liquid that absorbs the foreign matter from the air stream.

The separation of dissolved salts and other compounds from liquid streams can be accomplished by adding chemicals which precipitate an insoluble form of the compound, by evaporating the liquid and leaving a concentrated residue, or by using a material that has an affinity for the compound. Finely divided particles can be removed by adding flocculating agents which cause coagulation and settling of larger agglomerates. Many of these techniques have been employed to treat water prior to its use by cities and industries, but seldom have these techniques been applied to waste water treatment. Water softening units are common in homes and in treatment of major water supplies to remove dissolved salts. This salt removal can just as easily be performed prior to discharge of waste water.

Use of solid materials, such as activated carbon, that absorb gases, odors, or refractory organic liquids is a relatively new and advanced treatment concept. There are novel processes, such as ion exchange, membrane processes, evaporation, and radiation that have not yet been applied to large scale waste treatment. However, the use of the magnetic properties of iron to separate scrap metal is commonplace.

Conversion

The two basic methods used to convert waste to a stable or more suitable form for disposal are oxidation and biological stabilization. Some gaseous, liquid, and solid wastes can be incinerated under proper conditions and converted to carbon dioxide, water, and ash. In many cases, the method or the equipment limits the proper conditions for burning, and air pollutants such as carbon monoxide, organic fumes, particulates, and nitrogen oxides are formed. When there is improper combustion of solid waste, the residue will contain unburned organic matter which presents another disposal problem.

Biological treatment of waste water is another common method for removing wastes from water. This process is equivalent to disposing of solid wastes by natural degradation in landfill or composting. Biological treatment of air streams is currently a practical process. The natural aerobic and anaerobic assimilation of waste was discussed in Chapter 5. Processes for biological waste water or solid waste treatment attempt to provide ideal conditions for microorganisms to convert waste products to neutral forms.

Only organic matter is converted by biological processes. Microorganisms do not demonstrate particular skill at consuming cars, concrete, and other inorganic materials. The technical problems associated with biological waste treatment include the production of facilities that reduce waste to "bite sized" particles for the microorganisms, and provision of enough oxygen, proper temperature, moisture, and trace elements that the organisms require for optimum growth. A major problem with biological waste treatment is preventing buildups of toxic concentrations of waste that would kill the biomass. Chromates from metal plating are well known for causing kills in biological treatment plants. In oil refinery waste treatment plants, phenols, which are used as a disinfection agent, are present in high concentrations. Biological treatment of such wastes is accomplished by slowly acclimating the organisms to phenol and controlling the concentration of phenols in the waste water. Several processes utilize biological treatment, some of these are illustrated in Figures 8–2 through 8–5. Biological treatment is not the ultimate solution since biological processes create bacterial cell material.

Microorganisms are no different from other living organisms. If they are not fed a steady and balanced diet, they do not function to capacity. For example, municipal waste water is not uniform in composition, concentration or quantity. In sewer systems that also collect storm water, the organic content of sewage can be greatly diluted after a storm. Industries which operate only on weekdays will produce little waste on weekends. Food processing wastes tend to be seasonal, with several months of round-the-clock operation followed by shut-down for the rest of the year. It is a challenge to maintain a healthy culture that can degrade wastes with varied concentrations. In many cases, such fluctuations in waste output result in poor treatment.

Disinfection

One of the earliest environmental quality problems was the transmission of disease via wastes. One solution was to avoid waste contaminated water and

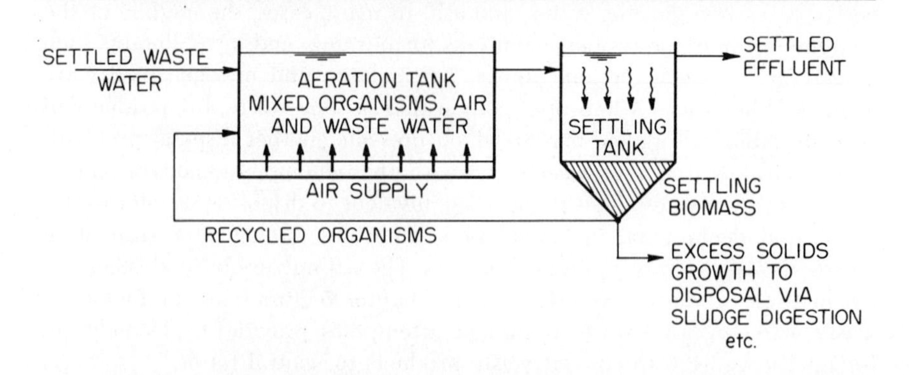

Figure 8–2. Typical Activated Sludge Type

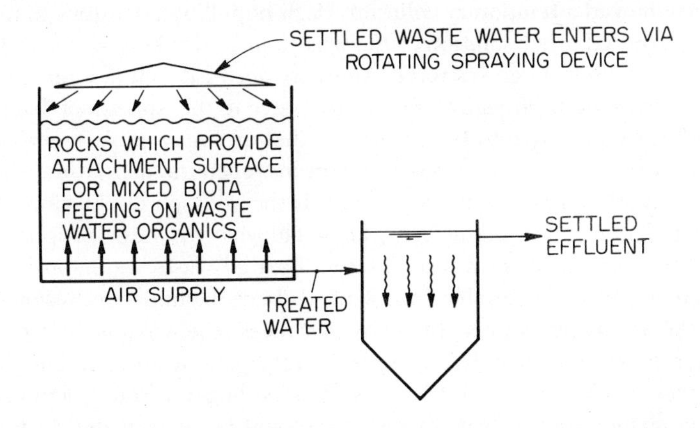

Figure 8–3. Typical Trickling Filter

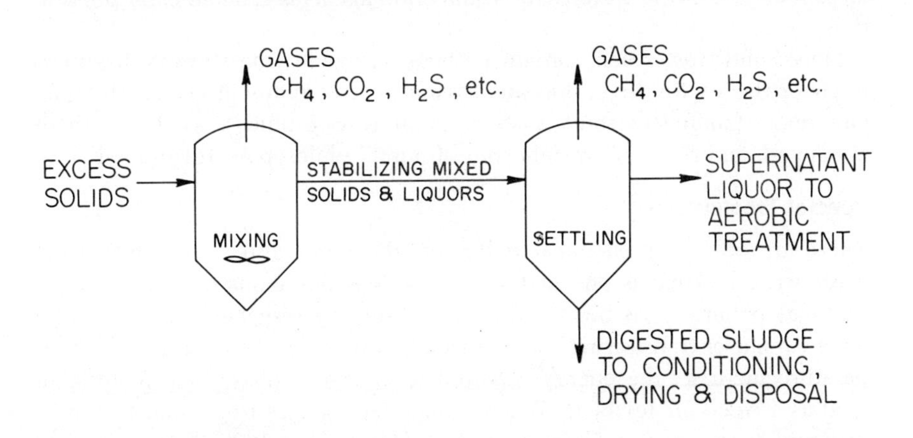

Figure 8–4. Typical Mixed Digester and Settling

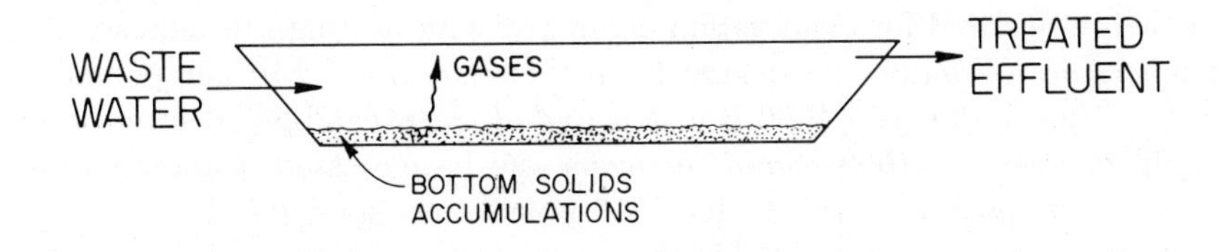

Figure 8–5. Typical Pond Treatment

air. By transporting waste in one direction and seeking water in the other, man achieved a temporary solution. High population densities in urban areas soon required other solutions. One was to filter drinking water through sand beds to remove large bacteria. Another more extreme action, practiced in Japan, was to wear gauze masks to prevent the spread of bacteria. Still another solution was to hold water so that an unfavorable environment was created for water-borne disease organisms, encouraging natural die-off.

The most common water treatment is the addition of chlorine compounds to kill pathogenic organisms. Adding chlorine to water supplies is now a routine practice in the United States. Treated waste waters are also chlorinated but some concern has been voiced about the impact of residual chlorine on the receiving waters. In many instances, the volume of treated waste water is greater than the volume of receiving waters, and high chlorine content can kill stream biota as well as pathogens. Thus chemicals such as ozone, which do not leave a residue, should be investigated further for use in waste disinfection.

Other methods of disinfection use heat, light, other chemicals, and radioactivity. Each of these has been demonstrated in special applications, but the large scale use of these alternate disinfection agents is economically prohibitive.

Since solid wastes can contain pathogenic agents, the strategy has been to cover them with dirt in fills and assume that vermin will not reach them. Our understanding of solid waste disposal is as primitive as the methods employed, and the "out-of-sight, out-of-mind" philosophy still prevails.

Special Treatment

There are other non-chemical pollutants that must also be removed from waste-streams. Heat is one of the most serious pollutants of waste waters. Thermal pollution has been controlled merely by dispersing the heat in a large volume or transferring energy to another medium. To minimize the temperature increase, the energy required to change a phase of a material is used as a reservoir for heat. When a material changes from solid to liquid, or liquid to gas, energy is required. It requires about 1000 British Thermal Units (Btu) to convert one pound of liquid water to steam. The water will be the same temperature either as a liquid or as a gas during this transition. If this process is reversed, 1000 Btu will be released. The advantage of evaporating rather than heating a liquid for heat dissipation is the difference in energy required for evaporation compared with heating. To increase the temperature of water one degree Fahrenheit requires 1 Btu per pound of water. Thus to dissipate 1000 Btu, *1 pound of water* can be converted from liquid to steam or *1000 pounds of water* can be increased in temperature by 1°F, 100 pounds increased 10°F, 10 pounds increased 100°F, etc.

Waste heat can be dissipated in the air or on the land. In any heat transfer process heat can be mixed directly with the cooling media, or it can be transferred to the media via the boundary surface. Normally, pipes are used to separate the hot fluid from the fluid to be cooled. Figure 8–6 illustrates several devices that can be used for heat dissipation. When hot water is released to the air it creates fog since the evaporated water can condense in the atmosphere. Dry cooling towers avoid this problem by separating the air and the water.

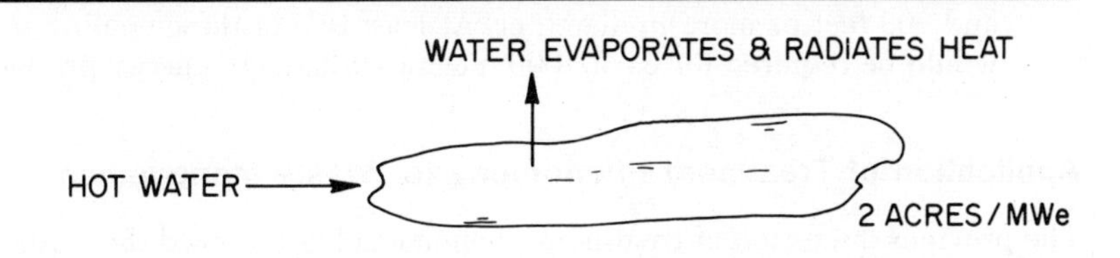

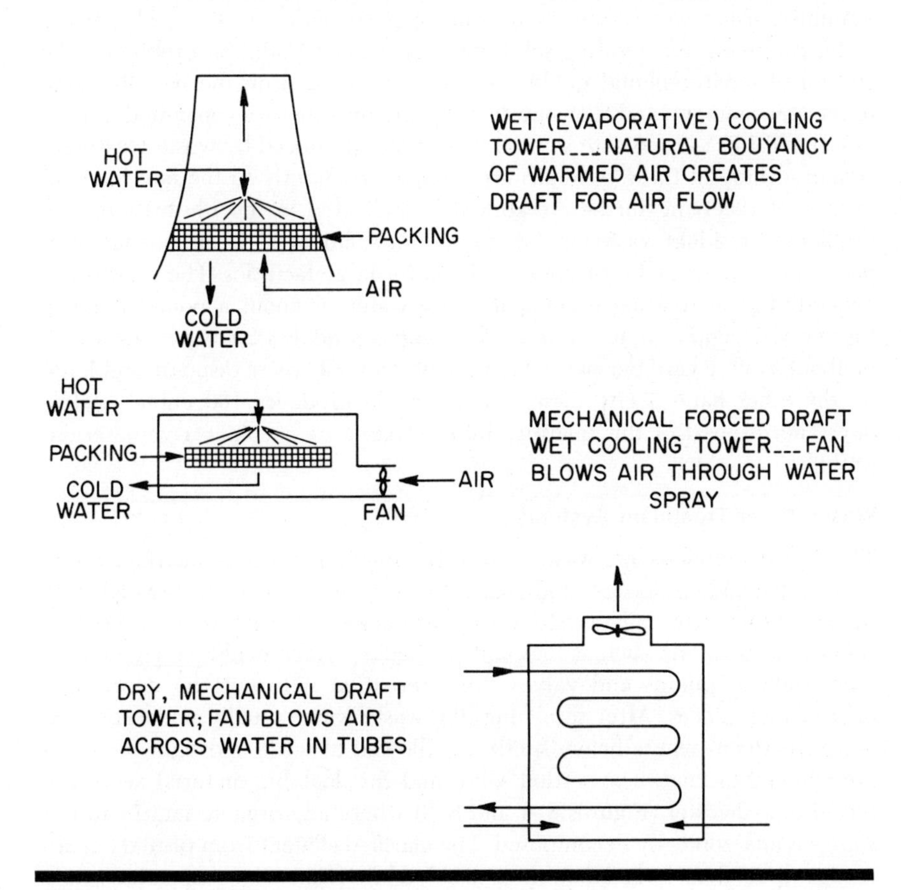

Figure 8–6. Water Cooling Processes

The most significant technical problem encountered in the control of thermal pollution is the sheer size of the cooling facilities. In a river flowing at the rate of 2000 cubic feet per second, one large power generating plant (1000 megawatt hours) can raise the temperature by 15°F. This is more water than is used daily by a million people:

> If a cooling pond were used, about 2000 acres of surface would be required; if cooling towers are used, they can be as high as 500 feet and 300 feet or more in diameter. At least two of these cooling towers would be required for each 1000 megawatt hours of energy produced.[1]

Application of Treatment Phenomena for Waste Management

The previous discussion of treatment phenomena has followed the traditional but fragmented approach of focusing on each pollutant independent of all others. Such a philosophy can create more problems than it can solve. For examples, using wet scrubbers to control particulates in air creates waste water problems; incinerating solid waste creates air pollution problems. The control of environmental quality needs to be focused on the net discharge to the environment, and to insure that all contaminants are controlled. Figure 8–7 and Table 8–2 indicate some of the wastes produced by waste treatment systems. Many of these residuals are discharged directly to the environment instead of receiving further treatment as indicated in the illustration. The problem of residual waste can be ignored for small waste treatment facilities but can become a major disposal problem for large facilities. The solid waste generated from treating municipal waste waters is about a pound per day for every 5 people. A town of 10,000 people produces 2000 pounds a day, or about a cubic yard per day. This amount does not create disposal problems, on the other hand a city of a million people produces 100 cubic yards a day. The disposal of this quantity in an urban area becomes a very serious problem.

Waste Water Treatment Systems

The major emphasis on waste water treatment has been the removal of oxygen-demanding wastes, and many types of equipment are available to remove BOD. Normally waste waters are screened prior to treatment to remove large debris, such as logs and car bodies, which could damage equipment such as pumps and valves. An alternative is to cut the debris into quarter-inch pieces. After screening, the waste water usually undergoes *primary* treatment where heavy solids are allowed to settle and light materials are allowed to float. The settled solids and the floatable material are transferred to a digester (Figures 8–4 and 8–8) where any organic matter in this *sludge* is anaerobically decomposed. The clarified effluent from primary treatment still contains large amounts of dissolved BOD plus fine suspended solids. The removal of this material can be accomplished by biological treatment in which engineered structures are designed to provide optimum conditions

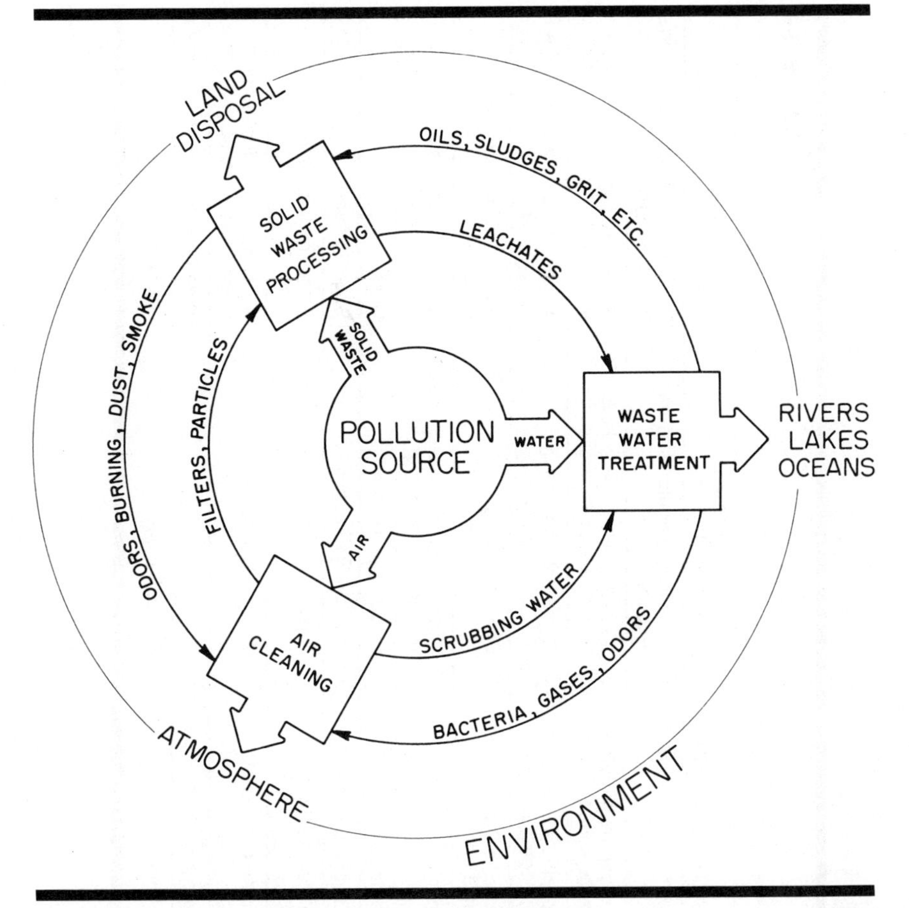

Figure 8–7. Generation of Wastes by Waste Treatment Systems

for biological decomposition. The effluent from secondary treatment (Figures 8–3 and 8–4) still contains a small amount of degradable organic material which also can be removed but at increasing expense. In practice, common treatment procedures do not remove any other types of pollutants from waste-streams. In fact, the use of water by municipalities contributes wastes to receiving waters after treatment. The levels are shown in Table 8–3.

Recognizing that waste waters contain undesirable materials other than oxygen-demanding waste has created the concept of renovation and reuse of water. It should be mentioned that there are alternate methods of waste water disposal for single family units, small towns, and remote hotels and institutions. These include septic tanks, preconstructed treatment systems, and lagoons. Each of these systems employs a settling and a biological step. Systems also exist which employ chemicals to destroy undesired pollutants. Chemical toilets are used in boats, campers, trailers, and airplanes, and they are more expensive than the other processes discussed above.

TABLE 8–2

Selectivity of Waste Treatment

TYPE OF WASTE	TREATMENT	WASTES REMOVED	MATERIALS ADDED OR CREATED	WASTES THAT ARE UNAFFECTED
Waste Water	Primary	Solids	Chemical coagulants in trace amounts	Dissolved materials, fine solids, heat pathogens
	Secondary	Oxygen demanding waste, solids	Bacteria cells recycled oxygen, nutrients if necessary	Refractory organics, dissolved salts, heat
	Tertiary	Specific salts, or refractory compounds	Chemicals, heat	Those not specifically selected for removal
	Chlorination	None, but pathogens killed	Chlorine, tastes	Solids and non-oxidable materials
	Cooling	Heat	Water vapor, chromates, corrosion inhibitors, algicides	Only heat removed
Air	Particulate Removal	Particles	Water or Solvent	Gases
	Gas Cleaning	Selected Gases	Heat, moisture, gas	Specific gases
Solid Wastes	Incineration	All combustible	Gases, particulates acid gas (HCl)	Glass, metals, stone
	Composting	Organic material	Water, CO_2, nutrients	Refractory organics, metal, glass, stone, plastics
	Reclamation	Selective separation		Mixed refuse difficult to process
	Landfill	All buried	CH_4, CO_2, odors	Glass, stone

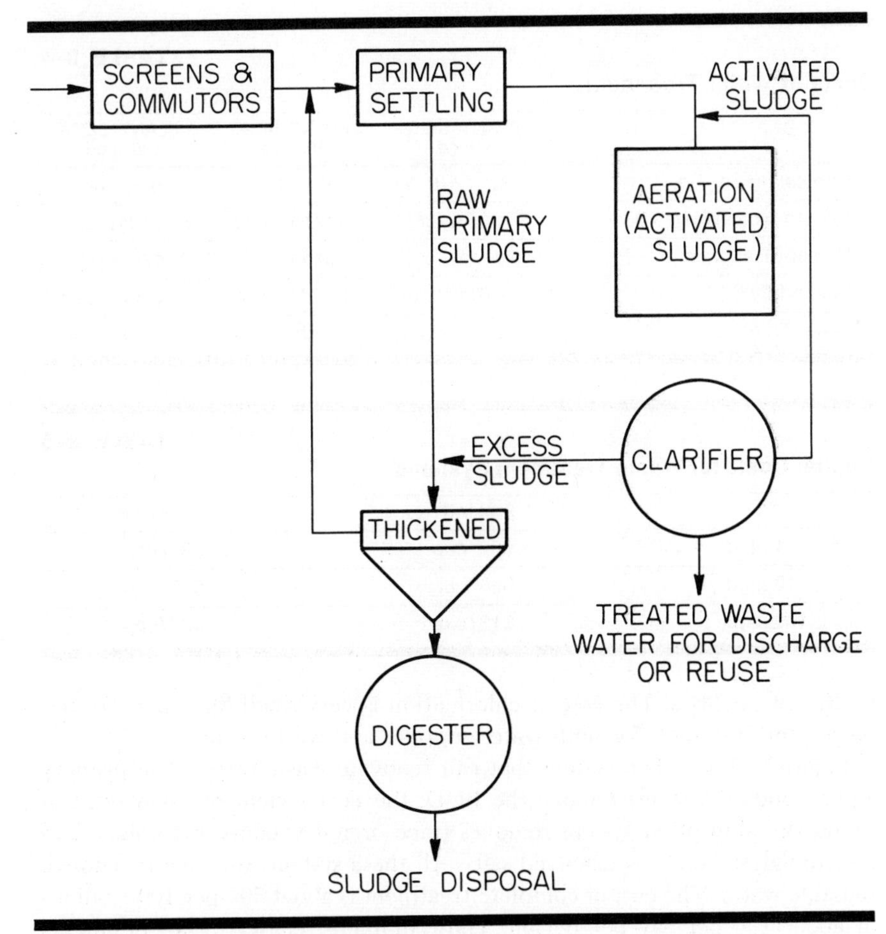

Figure 8–8. Sludge Generation in Biological Treatment

The costs for water treatment, as well as conventional primary and secondary waste treatment are shown in Table 8–4, and the initial capital costs for waste water treatment systems are indicated in Table 8–5.

The cost of single home units will vary from $500 to $1000, therefore, major savings can result from households close together that use central

TABLE 8–3

Average Chemical Content of Waste Treatment Plant Effluent (pounds per day per 1000 people)

Total Dissolved Salts	220
Phosphorus	20
Total Nitrogen	23

TABLE 8–4

Costs of Water Treatment

	PLANT SIZE 1 mgd	PLANT SIZE 10 mgd	PLANT SIZE 100 mgd
Population Served	6×10^3	6×10^4	6×10^5
Cost to Move Waste 100 Mi.	10¢/day/cap	3¢/day/cap	1¢/day/cap
Primary Treatment	2¢/day/cap	1¢/day/cap	½¢/day/cap
Secondary	3¢/day/cap	2¢/day/cap	1¢/day/cap
Water Treat	.7¢/day/cap	.6¢/day/cap	.5¢/day/cap

TABLE 8–5

Capital Costs for Waste Treatment Systems

	PRIMARY	SECONDARY
1 mgd	$52/cap	$82/cap
10 mgd	$25/cap	$48/cap
100 mgd	$13/cap	$20/cap

treatment facilities. The cost of chlorination is very small (0.1¢ per day per person for 100 mgd) for both water supplies and waste water.

Figure 8–9 depicts a system that can renovate waste water. The primary and secondary systems remove the BOD, the next system removes residual solids, the absorption system removes trace organics, odors and colors, and electrodialysis removes dissolved salts. All these systems produce reclaimed, reusable water. The cost of complete treatment is about 50¢ per 1000 gallons or about 10¢ per day per person. Thus complete removal of all pollutants can be achieved for large populations relatively inexpensively.

A detailed description of the limitations and potentials of waste treatment processes is given by the Advanced Waste Treatment Reports issued by the Environmental Protection Agency.

Investment costs for cooling facilities range from $7 per kilowatt to $30 per kilowatt for induced-draft, wet types. Using 10 to 15 percent fixed charge, the annual operating and maintenance costs would be comparable to the fixed charges for the cooling facility. For a cooling water flow of 0.5 gallons per minute per kilowatt (500,000 gallons per minute per 1000 megawatts), and an annual total cost of cooling of $2 to $10 per kilowatt, the cost of cooling thermal discharges from large power plants would range from 0.8¢ to 4¢ per 1000 gallons. Thus the cost of cooling water is less than the cost of removing organic oxygen demanding wastes.

In summary, the cost of water pollution elimination is not staggering and is on the order of 10¢ per day per person where central treatment facilities are used.

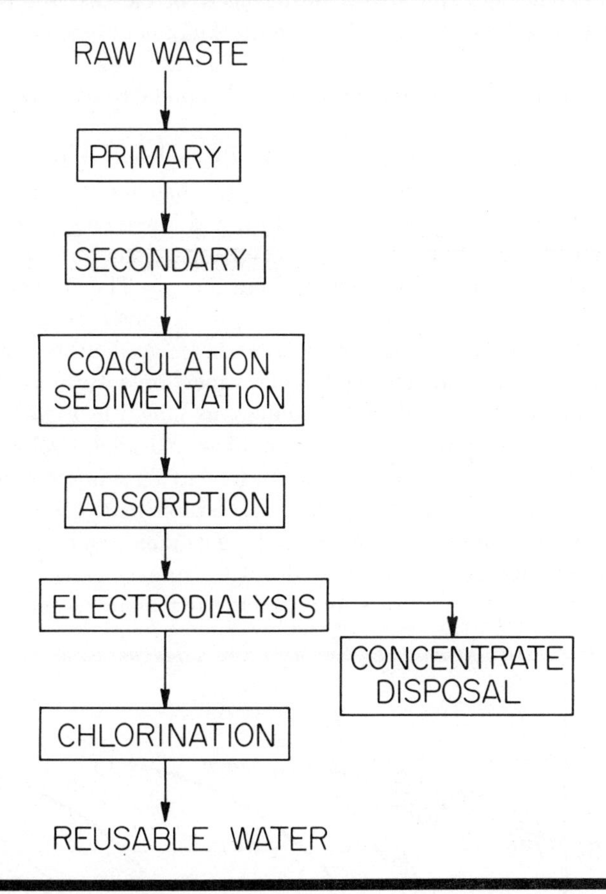

Figure 8–9. Tertiary Waste Water Treatment

Air pollution control must be separated into three separate categories: large stationary man-made sources, small mobile sources, and natural sources. While large man-made sources often have all gases collected and discharged through a single stack, there are cases where many sources exist within a single facility and separate control is required for each source. It is convenient to further separate air pollutants into particulates and gases to discuss control techniques.

Air Cleaning Systems

Gas cleaning is the most common technique used to control particulate air pollutant emissions from stationary sources. The removal of particles from their gaseous media can be accomplished using a number of methods including gravitational centrifugal, magnetic, inertial, thermal diffusion, Brownian diffusion, electrostatic precipitation, and diffusiophoresis forces. Selecting the proper gas cleaning equipment for a given source depends on the required collection efficiency, the nature of the gas to be cleaned, the particle properties, the availability of space to locate the equipment, and economic considerations. Important particle characteristics include size, distribution, density, hygroscopicity, corrosiveness, electrical conductivity, flammability, and toxicity.

Particle collection efficiency is greatly influenced by particle size, as shown in Figure 8–10. Thus particles larger than about 20 microns are easily removed by cyclones, inertial collectors, and low-energy wet scrubbers. Cyclones are devices that utilize the centrifugal force created by a spinning gas stream to separate the particles from the gas. The conventional cyclone, shown in Figure 8–11, is relatively simple to construct and has no moving parts. The rotational motion of the gas can be applied in a number of ways including a tangential gas inlet, inlet vanes, or a turbine. The tangential inlet is most common on single cyclones, as shown in Figure 8–11, whereas the axial inlet cyclone is used with multiple (parallel) cyclones.

Wet scrubbers use a liquid (usually water) in a gas-liquid contacting device to transfer the particles from their suspension in a gas to a liquid phase suspension or solution. To maximize the gas-liquid contact surface area, the liquid is applied in the form of droplets (sprays) or bubbles (foam). The advantages of wet collectors include no particulate re-entrainment, ability

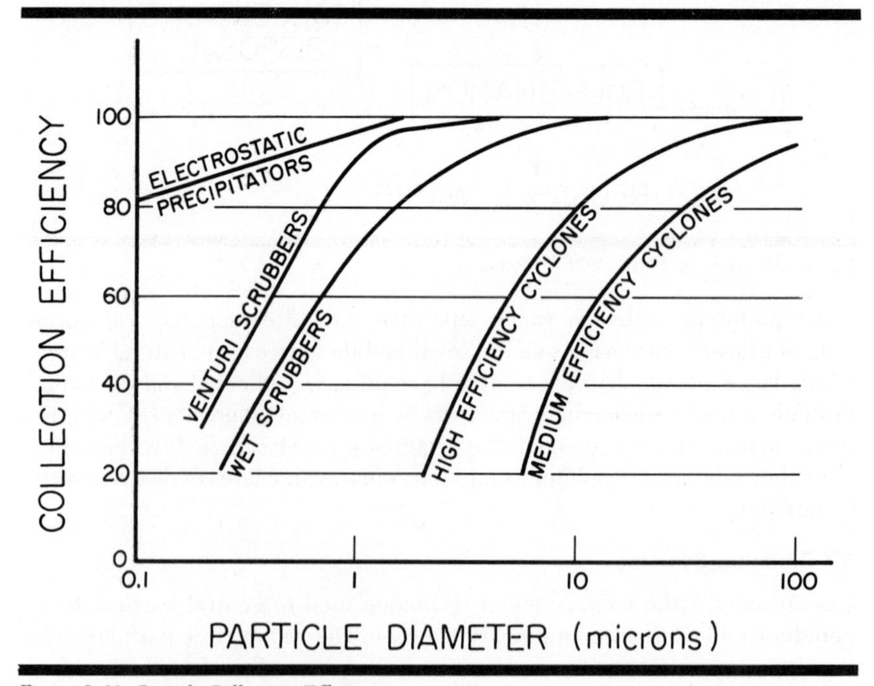

Figure 8–10. Particle Collection Efficiency

From AIR POLLUTION CONTROL: GUIDEBOOK FOR MANAGEMENT, ed. by A. T. Rossano, Environmental Resource Associates, Inc., 1969.

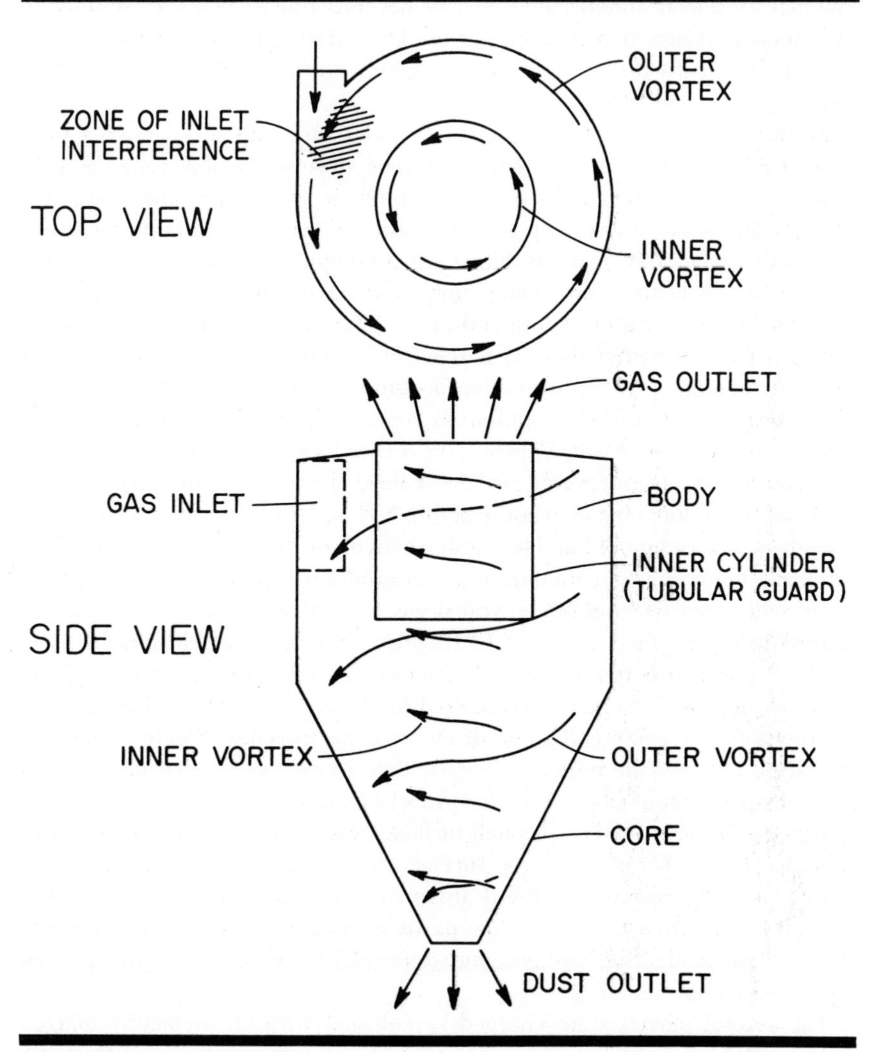

Figure 8–11. Cyclose Separator

From CONTROL TECHNOLOGY FOR PARTICULATE AIR POLLUTANTS, National Air Pollution Control Administration, Pub. AP–51, 1969.

to handle high humidity gases, small space requirements, ability to collect particulate and gaseous pollutants simultaneously, and cooling of high temperature gases. The disadvantages are the difficulty of removing the particles from the scrubbing liquid, the high relative humidity of the washed air, causing steam plume problems, and the problem of disposing of the waste scrubbing liquid. There are many types of wet scrubbers including spray towers, cyclonic scrubbers, floating beds, venturi scrubbers, induced draft scrubbers, and wetted filter scrubbers. A floating bed scrubber consists of low density plastic spheres about 1.5 inches in diameter that are continually in motion between two retaining grids. The rotating motion of the spheres continually cleans the particles off the sphere surface which prevents plugging.

Filtration through fabric is one of the oldest methods for removing solid particulates. A filter is essentially a porous structure which removes particulate matter from a fluid passing through it. In general there are two types of filters; fibrous or deep-bed filters (low efficiency) and fabric or paper filters (high efficiency). Fibrous filters are commonly used to filter air passing through home heating or cooling units and have low pressure drops (less than 0.5 inches of water). Cloth and paper filters are capable of high collection efficiencies (greater than 95 percent by weight) even for small particles (0.1 micron diameter) and have high pressure drops (1 to 6 inches of water). The fabric filters used in air pollution control are usually arranged as bags in a baghouse, as shown in Figure 8–12. Filter bags are cleaned periodically via shaking, air jets, or reverse air flow. Fabric filters are composed of various materials including cotton, wool, Dacron, Nylon, Orlon, Nesmex, Teflon, and fiberglass. The design of bag filter systems involves the use of the gas-to-cloth ratio: the cubic feet per minute of gas filtered per square foot of filter area, expressed as feet per minute. Typical gas to cloth ratios are 3 ft./min. for intermittently cleaned bags and 12 ft./min. for reverse jet cleaning.

Electrostatic precipitators are frequently used in large industrial plants (such as pulp mills, cement plants, coal fired power plants, steel mills, etc.) for the high efficiency collection of airborne particulates. Electrostatic precipitation involves the use of an electric field to remove electrically charged particles from their gaseous media. An electrostatic precipitator has a discharge electrode (negative) of small surface area, such as a wire, and a collection electrode (positive) of large surface area, such as a plate, as shown in Figure 8–13. There are essentially three steps in electrostatic precipitation: placing an electrical charge on the particles, migration of the charged particles to the collecting surface, and removal of the particles precipitated on the collection electrode.

The aerosol particles are charged by collision with air molecules ionized at the discharge electrode. The precipitated particles are removed from the collecting surface by rapping, scraping, vibration, or washing. There are two general types of electrostatic precipitators; low voltage (two stage) and high voltage (single stage). The low voltage precipitator consists of an ionizing stage followed by a collecting stage and is commonly used in air conditioning systems, such as the units sold for home use. High voltage or single stage electrostatic precipitators combine ionization and collection and are used in removing particles from highly concentrated emissions. Flat collection electrode (plate) high voltage electrostatic precipitators are commonly used in industrial plants.

There are essentially four methods for removing gaseous pollutants: (1) absorbing the pollutant into a liquid, (2) adsorbing the pollutant onto the

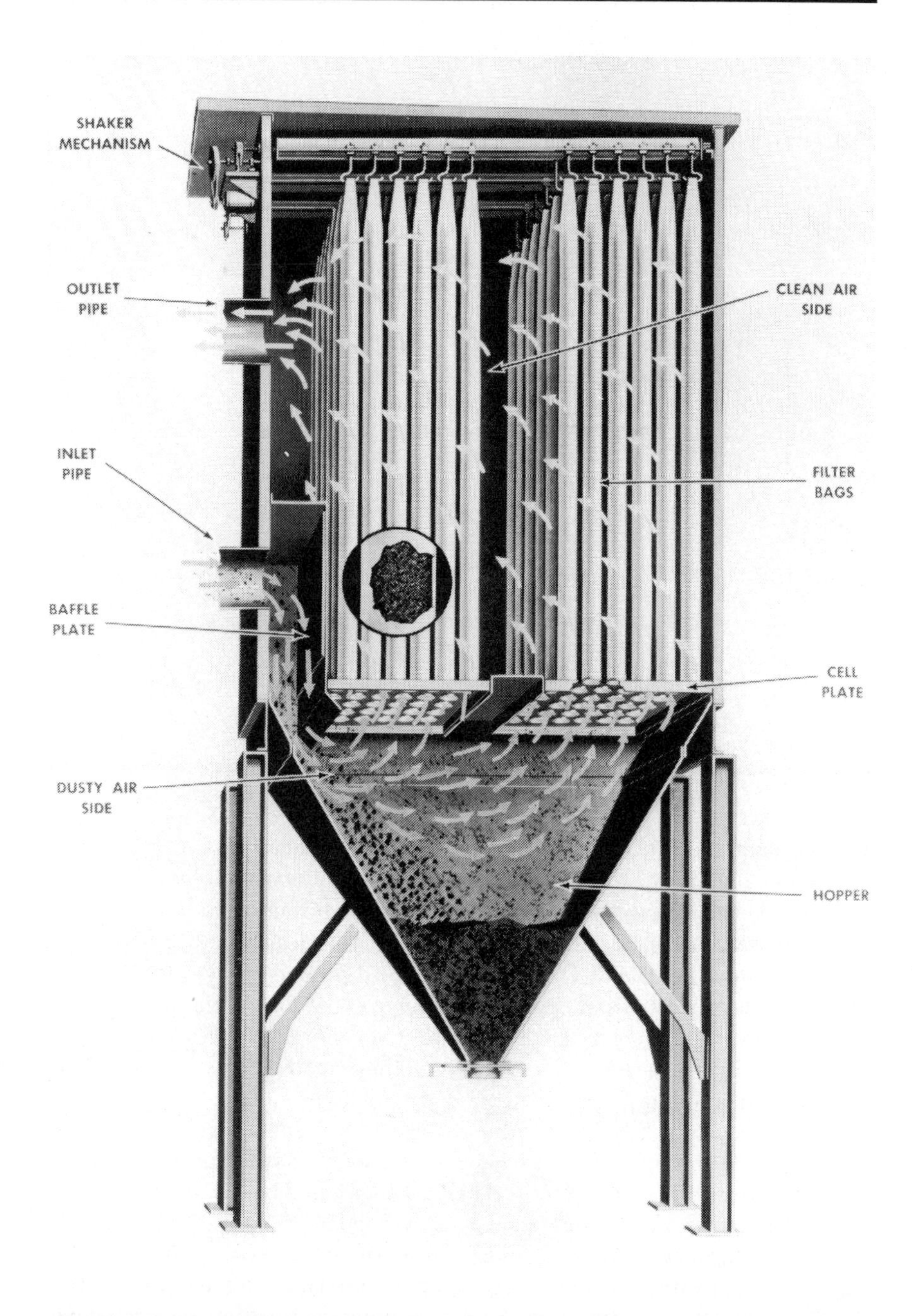

Figure 8–12. Wheelabrator Dustube Dust Collector
Courtesy Wheelabrator Air Pollution Control, Portland, Oregon.

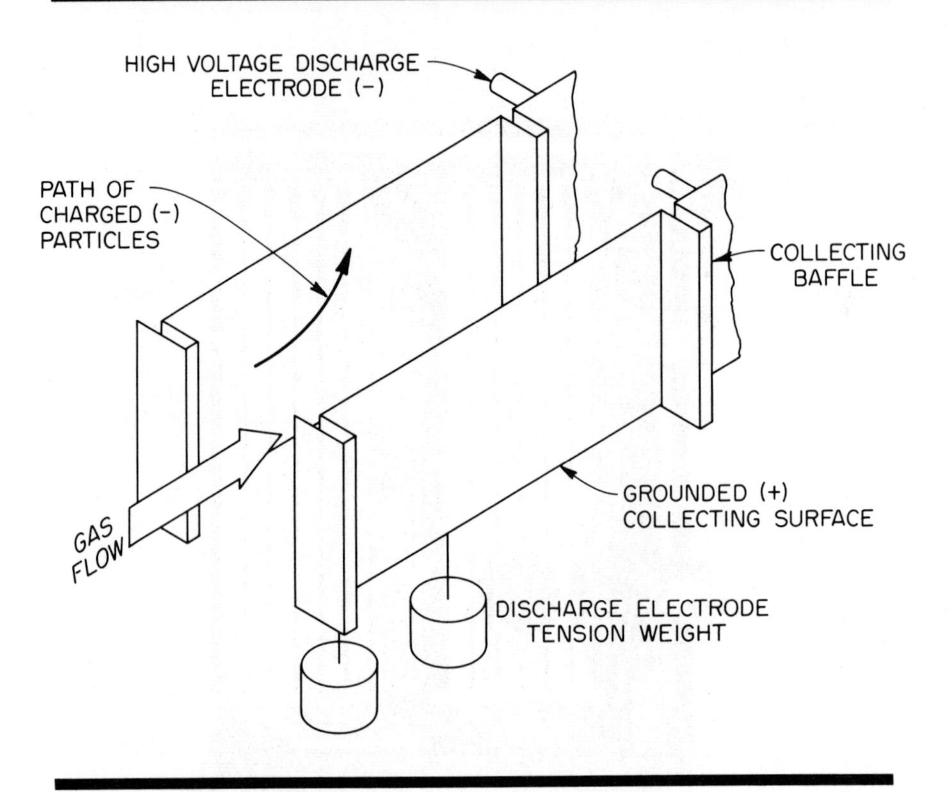

Figure 8–13. Electrostatic Precipitator

surface of a solid, (3) condensing the gaseous pollutant into a liquid, and (4) chemically changing the pollutant into a non-pollutant substance. Absorption has been successfully used to remove sulfur dioxide from smelter gases. The dimethylaniline process developed by the American Smelting and Refining Company (the ASARCO process) is used to produce pure sulfur dioxide. The gaseous exhausts from copper smelting are first passed through an electrostatic precipitator to remove particles, then the gases are scrubbed with pure dimethylaniline (DMA). After scrubbing with DMA, the gases are washed with a sodium bicarbonate solution and dilute sulfuric acid to remove the remaining SO_2 and DMA. The DMA solution containing the absorbed SO_2 is passed through a heat exchanger to a steam distillation tower which removes the sulfur dioxide from the DMA solution. The SO_2 is then passed through a cooler, drying tower, and finally condensed into a liquid.

In air pollution control adsorption is primarily used to collect hydrocarbon vapors on activated carbon. Processes that discharge organic vapors that can be controlled by adsorption are dry cleaning, degreasing, paint spraying, and gasoline vapor emissions. When the adsorbent becomes saturated with the pollutant it must either be replaced or regenerated. Regeneration can be accomplished either by raising the temperature of the adsorbent until the vapor pressure of the adsorbed gas exceeds the atmospheric pressure or by purging the adsorbent with steam, as shown in Figure 8–14. Steam is passed through the adsorbent bed, and the mixture of steam and hydrocarbons is condensed and separated in a decanter. By using two adsorbent beds one bed can undergo regeneration while the other is in operation.

The process of condensing an organic vapor into a liquid is used to control hydrocarbon emissions. Many organic compounds have boiling points greater than 250°F, and readily condense even when they are not highly concentrated. However, the control of hydrocarbon emissions by condensation is limited by the equilibrium vapor pressure of the organic material. For example, at 32°F toluene has an equilibrium vapor pressure of 6 millimeters mercury, and thus at atmospheric pressure (760mm Hg) a gas stream will still contain about 8000 ppm toluene. Therefore, to achieve greater collection efficiency, condensers are sometimes followed by adsorbers or afterburners.

Chemical reaction of gaseous air pollutants involves mainly the oxidation of carbon monoxide and hydrocarbons by combustion. The oxidation process can be self-sustaining if the concentration of combustible pollutants is sufficient. To achieve complete combustion, the proper combination of temperature, combustion time, and fuel-oxygen mixing (turbulence) is necessary. Combustion can be accomplished by direct-flame incineration (afterburners) or by catalytic oxidation. In order for a flame to be self-sustaining, the heat required is about 50 Btu per standard cubic foot of gas. This energy can come either from the pollutant gas or from an auxiliary fuel. Burning time (residence time) and temperature are important to the performance of an incinerator. For example, open burning such as piles of trash, and forest fires, emit large quantities of smoke from incomplete burning caused by low temperatures and insufficient reaction time. A schematic diagram of a multijet direct flame afterburner which uses polluted air as the source of oxygen for the flame is shown in Figure 8–15. The incineration chamber serves to keep the gases hot long enough for the gases to oxidize completely. Properly designed and operated direct flame afterburners usually remove more than 95 percent of the organic vapor. If the combustion reaction is halted by low temperature, insufficient residence time, poor mixing, or insufficient oxygen, carbon monoxide, aldehydes and other products of incomplete combustion may be produced.

Catalytic combustion can be used to oxidize combustible pollutants in emission streams which do not contain catalyst poisons. Catalytic incinerators employ a solid surface upon which the oxidation reaction can occur at a temperature somewhat lower than that needed for direct flame combustion. Catalytic incinerators have energy requirements of about 4 to 10 Btu per cubic foot per second, and thus have the advantage of lower fuel costs than direct flame afterburners.

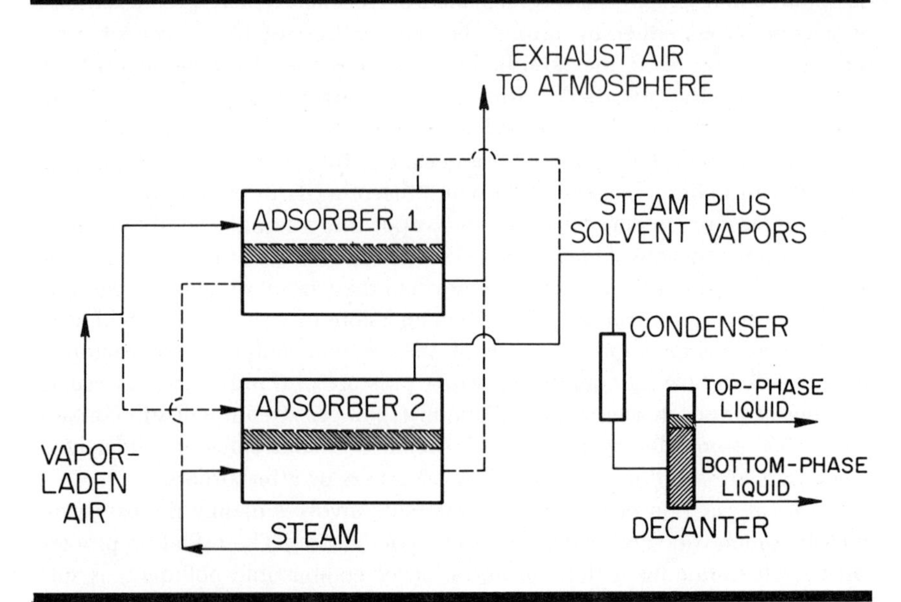

Figure 8-14. Air Cleaning by Adsorption

From CONTROL TECHNOLOGY FOR HYDROCARBONS AND ORGANIC SOLVENT EMISSIONS FROM STATIONARY SOURCES. National Air Pollution Control Administration, Pub. No. AP-68, March 1970.

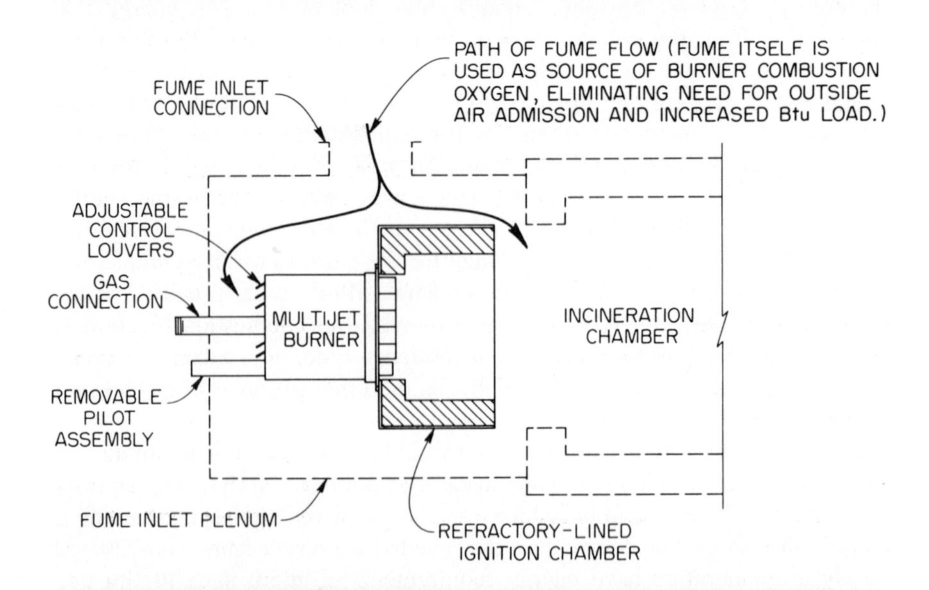

Figure 8-15. Multijet Afterburner

Courtesy of Hirt Combustion Engineers.

Motor vehicles are the major source of carbon monoxide, nitrogen oxides, and hydrocarbons in the United States. Emissions from a gasoline-powered vehicle without any emission control systems come from the engine exhaust, crankcase, carburetor, and gas tank. Exhaust is the source of almost all the CO and NO_x and more than 50 percent of the hydrocarbons, as shown in Figure 8–16. Hydrocarbon emissions from the carburetor are caused by the evaporation of gasoline after the engine is turned off (hot soak time period). Evaporation from the gas tank occurs primarily during hot weather. Emissions from the crankcase are caused by exhaust gases which pass the piston rings, then escape to the atmosphere through the road draft tube on the crankcase ventilation cap.

Crankcase emissions are controlled with positive crankcase ventilation systems which pass air through the crankcase and into the engine intake manifold to be burned, as shown in Figure 8–17. Such systems were first used in 1961 model autos in California and in 1963 models nationwide. Exhaust emission controls for hydrocarbons HC and carbon monoxide CO became effective for 1966 models in California and 1968 models nationwide. In general, two approaches are used to control HC and CO emission in these cars: air injection into the exhaust manifold, and use of a lean fuel mixture plus retarded spark. The air injection system reduces the CO and HC emissions

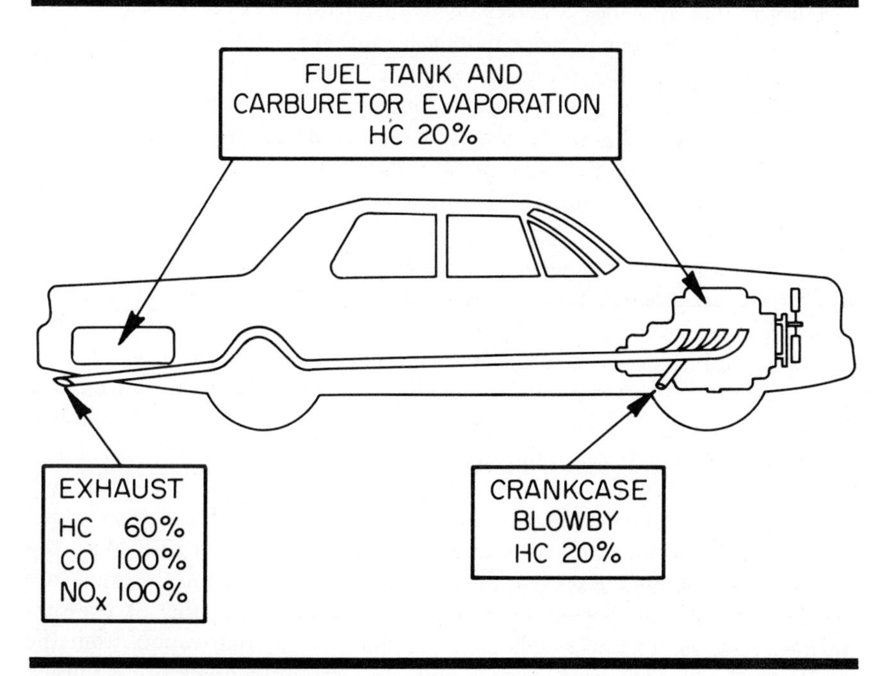

Figure 8–16. Automobile Emissions

From CLEANING OUR ENVIRONMENT: A CHEMICAL BASIS FOR ACTION. American Chemical Society, 1969.

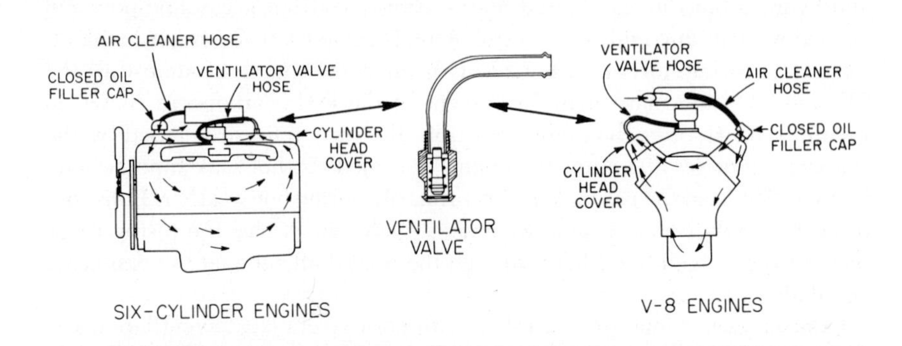

Figure 8-17. Positive Crankcase Ventilation in Automobile Engines
From CONTROL TECHNOLOGY FOR CARBON MONOXIDE, OXIDES OF NITROGEN AND HYDROCARBONS FROM MOBILE SOURCES. National Air Pollution Control Administration, Pub. No. AP-66, 1970.

by taking in air through an air filter, the air supply pump, the air manifolds, and through air injection nozzles located at each cylinder exhaust port. The oxygen in the air oxidizes the unburned hydrocarbons and CO into carbon dioxide and water vapor. The engine modification approach (used on most 1969 through 1972 autos produced in the United States) includes a modified carburetor that provides a relatively lean air-fuel mixture together with higher engine idle speeds and a modified ignition system which retards the spark during idling. Retarding the ignition timing at idle tends to reduce CO and HC emissions in two ways: by increasing exhaust gas temperatures and by requiring a slightly larger throttle opening (increased gasoline and air flow) to obtain a desired idle speed. At larger throttle openings, emissions are reduced by better mixing of the fuel and air, resulting in better combustion.

Gasoline evaporative controls were required on 1970 models in California and on 1971 models nationwide. Two evaporative control systems are presently in use: adsorption by activated carbon, and crankcase vapor storage. The activated carbon bed system adsorbs gasoline vapors emitted from the gas tank and the carburetor until they can be injected back into the carburetor for combustion in the engine. When the engine stops operating, gasoline vapors from the carburetor and the fuel tank are vented through a canister filled with activated carbon. When the engine is started, air passes through the activated carbon bed and removes the adsorbed gasoline vapors which are then injected into the carburetor to be burned in the engine. With the crankcase vapor storage system, the decreasing temperature in the crankcase during the hot soak period after engine shutdown causes the air pressure in the crankcase to lower sufficiently to draw in gasoline vapors from the carburetor and gas tank. When the engine is started the crankcase is purged of vapor by means of the positive crankcase ventilation system.

Nitrogen oxide emissions from motor vehicles have increased because of efforts to reduce emissions of HC and CO. At the present, autos contain no controls for NO_x emissions. A number of approaches for controlling NO_x emissions are available including exhaust gas recirculation, use of a rich air-to-fuel ratio with an exhaust manifold reactor to burn the remaining HC and CO, and catalytic reduction.

The cost of such controls is variable but ranges around a few dollars per cubic foot of gas per minute. Another example: the Kaiser Aluminum Refining Plant in Tacoma, Washington has a waste gas flow of about 700,000 cubic feet per minute. The cost of a wet wash scrubber for the installation was $3.50 per cubic foot per minute, or about $2,500,000. In some cases, by-products have made air pollution control more feasible, such as at Bethlehem Steel in Seattle. The installation of a bag filter house at that scrap iron reprocessing plant cost several million dollars. Maintenance requires replacing the fiberglass filter bags every few years at a cost of a few hundred thousand dollars. The recovered material is sold as a source of zinc and for other trace minerals. The whole installation process results in a system which nearly pays for itself. Finally on the individual level, an automobile produces about 50 cubic feet of exhaust per minute; the cost of about $50 for the control mechanism is within the general figure of $1 per cubic foot of effluent per minute for capital costs.

Solid Waste Disposal Systems

Solid waste disposal may be accomplished by many methods, including incineration, landfill, composting, salvage and reclamation, central grinding, dumping at sea, and hog feeding. Landfill may vary from dumping and burning to sanitary landfill operations. All methods of refuse disposal except landfill and dumping at sea leave a residue which must be disposed of by one of these two methods.

Incineration is only an intermediate step in solid waste disposal since the ash must be removed and large quantities of gases are discharged. Only organic materials can be incinerated; glass, metal, and other noncombustibles are not suitable for incineration. The incinerator combustion process is concerned with the chemical combination of organic matter and oxygen in the air. An incinerator consists of the primary chamber where preheating and combustion occur, a secondary chamber for combustion and expansion of gases and settling of fly ash, a stack to discharge the gases, and a feed and ash handling system.

The heating value of refuse is between 7000 and 8000 Btu per pound, dry and ash-free. The control of air and auxiliary fuel is critical for proper combustion. Improper combustion can release particulates and gases which create serious air pollution problems.

In one form or another, landfill has been the primary method of refuse disposal for centuries. This method has a disreputable image because to most people landfills are open dumps, emitting offensive odors, providing ideal breeding places for rodents and insects, and blanketing the adjacent countryside with smoke and other combustion products when burned intentionally or unintentionally. The large volume of refuse-hauling traffic on streets leading to landfills is also objectionable, particularly if uncovered vehicles are permitted and if principal haul routes pass through residential areas.

It has been found that most of the health and nuisance hazards of landfilling can be substantially eliminated by using the sanitary landfill method, by rigid vector control, and by adopting and enforcing vehicle regulations that minimize the scattering of refuse on streets and highways. In sanitary landfilling, all refuse received is compacted and covered daily with earth or other inert granular material. This cover either prevents the propagation of rodents and insects in the fill or so limits their activity that they can be controlled and makes the completed fill more stable and useful. Completed sanitary landfills are suitable for parks, golf courses, parking lots, outdoor storage, or other uses requiring open areas in which settlement and surface irregularities can be tolerated. By using pilings and special construction, buildings can be placed on relatively shallow fills.

From the standpoint of disposal cost only, sanitary landfilling is the most economical disposal method available to most municipalities in the United States. However, the cost of hauling refuse from the collection area to the disposal site must be considered in an evaluation of various disposal methods. Recently, many cities have found that the added cost of hauling to distant landfill sites more than offsets the higher disposal cost of other methods utilizing facilities which could be located much nearer to the collection areas.

Composting is an alternative process by which the organic components of refuse are converted into a humus by controlled, biological oxidation. The compost has value as a conditioner for sand, clay, or other organically-poor soils.

The fundamental steps for aerobic composting include removal of noncombustibles, grinding, moving and placing for composting, turning or aerating, and regrinding, bagging or storing. Materials with a ratio of carbon to nitrogen greater than 50 percent are very slow to compost, and a moisture content of 40 to 65 percent is desirable for rapid composting. Maximum depths of 5 to 6 feet prevent compaction, but minimum depths of 4 feet are needed for insulation. The temperature will rise to more than 60°C 2 or 3 days after composting, thus killing pathogenic bacteria. Composting is completed usually in 12 to 20 days. It requires adequate bottom drainage and air passage through the site, and is not attractive in urban areas since it only changes the form of the waste and does not resolve the ultimate disposal problem.

Currently, the only salvage methods warranting consideration are those which do not interfere with the primary objective of disposal, are largely mechanized and require little labor, and recover items which have remained relatively stable in value and for which markets are reasonably assured in the future.

Tin cans constitute 80 to 90 percent, by weight, of the metal salvaged from refuse. The remainder is largely ferrous metal scrap. However, most steel mills cannot accept tin cans, even if they have been burned and detinned, as the small residual of tin has deleterious effects on furnaces and on finished products. The primary markets for salvaged tin cans are copper mines which use shredded cans in the ore extraction (precipitation) processes. Glass must be salvaged, if incineration is employed, as the melting point of glass can be reached in the combustion chamber. To achieve the maximum salvage value for glass it must be sorted into types, sizes, shapes, and colors. Broken glass has a limited salvage value for reuse in bottle manufacturing, but even this should be sorted according to color.

Central grinding and discharge of solid wastes to sanitary sewers has been suggested. To operate a grinding station, it is necessary to separate non-grindable from grindable materials which can be flushed into sewers. This requires sorting either by separate collection or hand sorting at the station. Research is currently in progress to develop pipeline transfer of solid waste with minimal grinding. Refuse could be ground and discharged in any convenient place, but would add additional flow to the sewers and additional hydraulic and solids loads to the treatment plant, thus necessitating the construction of additional treatment capacity to accommodate the increased flow.

Dumping solid wastes at sea has been practiced by a number of seacoast cities but has been discontinued because the residues invariably washed ashore and polluted beaches and other facilities. This practice is not a practical solution unless the refuse is compacted to a sufficient density or weighted to assure that it sinks. At times when the weather does not permit, barging has to be discontinued and refuse stored until operations can be resumed, thus leading to problems with rodents, odor control, and fire hazards, as well as other problems associated with refuse storage.

The cost of solid waste control is mainly transportation expense. Some localities have allocated more money to process and disposal than to transportation, but these are usually exceptions. Collection costs vary from $5 to $25 per ton of solid waste. Since no common reporting system exists for collection costs, many reported collection costs also represent transportation of collected waste to disposal sites. Thus the low reported values for collections are for small towns with nearby dumps and high values represent large urban areas with distant disposal sites. Current disposal costs are minimal for solid wastes. An open burning dump costs less than 25¢ per ton while a properly operated landfill may cost more than $1 per ton to operate. Incineration costs about $5 per ton, becomes attractive when transportation costs exceed this amount. The cost of air quality control for incinerators could double or triple this cost, however. Administrative costs can add about

$1 per ton. As with other environmental control costs, the full cost of control is usually avoided due to "subsidies" such as low cost public land, or inadequate treatment of waste prior to discharge.

Economies of Scale

Size of Treatment Systems

Operation size is a challenging problem in environmental quality control. Volumes of waste are usually so large that even the simplest treatment process requires major investments. Large size is not a necessary characteristic of treatment facilities, but many treatment facility designers use land-intensive rather than labor or capital-intensive devices. Solids separation employing gravity requires many more acres than separation with mechanical gravity devices. The same is true for biological treatment versus chemical treatment. If an organism requires hours to degrade waste and chemicals require seconds, the container for the reactions can be significantly reduced by using chemical treatment. Technology is currently available to reduce the size of waste processing equipment and to reduce the volumes of waste. The entire waste control problem requires *incentives* to apply control more than it requires the development of new concepts.

Transportation of Wastes

Transporting waste to central treatment facilities and transporting treated waste to locations that can accept them is a major activity of environmental quality control. Solid waste control currently devotes its major resources to collecting and moving wastes to disposal sites. Transportation is a major factor in solid waste management activities, yet solid waste hauling is not considered in the design and planning of transportation facilities. The demand for garbage trucks and landfill equipment is not great enough to stimulate vehicle designs for efficient solid waste collection. Instead most of these vehicles are makeshift adaptations of standard vehicles.

Another major transportation system for environmental control is the sewer system. Sewers represent a major investment in urban water quality control. In fact, in many areas, the collection of waste waters and their transport downstream represents the extent of waste treatment. Unless waste waters are collected in many regions, there is no waste treatment. Individual treatment systems for homes and small businesses are not effective and current technology does not offer any low cost solutions.

While collection of polluted air does not seem feasible, major industries are constructing large ductworks to collect their contaminated air. The metal refining industries have constructed hoods over their pot lines, and have enclosed large process areas to control air quality. The concept of attaching a hose to every car to collect exhaust gases may be laughable, but collection of vehicle exhaust by some means may be a necessity in the future if source control is not affected.

Ultimate Disposal to the Environment

Waste treatment or reducing sources of waste does not eliminate the need for discharging waste to the environment. In order to control the impact of waste on environmental quality, three alternative concepts are available. Waste can be released in the least damaging location in an attempt to improve local conditions at the expense of the total environment. The environment can actually be treated to improve the quality, or man can become desensitized and accept lower levels of environmental quality. A final alternative may be to strive for a closed cycle economy where someone's waste is another's basic input for production.

Dilution, Detention, Dispersion and Diversion

Several methods have been used to control environmental quality. They are low in cost, but do little to enhance overall environmental quality. Instead they improve local conditions at the expense of the total environment. The methods are:

Dispersion: The distribution of a waste over a wide area of the land or into a larger volume of air or water.

Dilution: The artificial augmentation of the volume of the environment used to assimilate waste.

Detention: The temporary holdup of production or release of discharges for later gradual release or for release at a more advantageous time.

Diversion: The transportation of waste to another location for discharge.

Dispersion is the most widely used of these four alternatives. For example, tall smokestacks increase the area over which gaseous materials are discharged; diffusers transport material out to receiving waters to achieve greater dilutions, and spreading liquid waste over land rather than discharging it to water achieves assimilation. There is a physical limit to the height of a smokestack or the length of a diffuser outfall. In one case, the initial outfall was a mile long and sewage discharge was objectionable since onshore currents washed the sludge back onto the beach. The diffuser was extended in progressive increments into deeper water until it reached a length of 7 miles, and there are still times when sludge is washed back onto the beach. Dispersive devices such as diffusers and smokestacks have definite advantages in selected conditions. As with all of these alternatives, each has its place and provides economic advantages under specified conditions.

Diluting waste discharges rather than treating them is another widely accepted method of environmental control. However, this method has undergone critical analysis since the cost of supplying additional volumes of the environment, in most cases water, is becoming prohibitive. Building dams to supply volumes of water to dilute wastes can be a misleading advantage. For example, if a river's flow measures 1000 cubic feet per second, an additional 1000 cubic feet per second must be provided artificially in order to dilute the waste by only a factor of 2. Examining the relative cost of providing 1000 cubic feet per second of water versus the cost of removal of half the waste discharge often leads to the conclusion that it is much cheaper to treat wastes than it is to augment the water flow. A more convincing argument against the use of dilution is demonstrated in the case of very large rivers such as the Columbia. The low flow of the Columbia River is approximately 75,000 cubic feet per second. To achieve a dilution factor of 2, another 75,000 cubic feet per second must be stored for the low flow period. If this alternative were to be chosen it would require the total storage of Grand Coulee Dam to provide enough water to dilute wastes by a factor of 2 for even 1 month. Applications where dilutions can be highly beneficial are in rivers where the water flow in the summer is reduced to almost zero. In this case, the availability of even small quantities of stored water can have large beneficial effects since the volume of water necessary to achieve dilution factors of 10 or 100 is relatively small.

Another interesting concept is the detention of waste until large volumes of water are naturally available to dilute it. For example, if a river has very low flow for short periods of time, it may be more economical to build a storage lagoon or holding tank to accumulate waste until high river flows are available and the total waste can be discharged. An interesting example of waste detention is practiced by the Crown Zellerbach pulp mill in Camas, Washington. The problem in this case is the growth of a slime bacteria called Sphaerotilus. Their growth presents a major problem in the Columbia River since they can become detached and foul fishing nets as well as create large piles of slime along shorelines. Research by the Crown Zellerbach scientists indicated that the growth can be halted by intermittent rather than continuous release of pulp mill wastes. In order to test this theory, a pond was constructed on an island in the middle of the Columbia River. It provided sufficient storage for 6 days' pulp mill output. The pulp mill effluent was released 1 day a week rather than continuously. This experiment did reduce the growth of Sphaerotilus in the Columbia River.

There are other industries which may be able to benefit by using detention. Seasonal operations, such as food processing, create their peak waste loads when river flows are low, creating severe treatment problems since large amounts of waste must be treated in relatively short periods. The alternative to treatment in these cases would be to create storage areas large enough to hold the waste until high river flows are available to assimilate them. One of the major problems arising from detention is that organic wastes decompose and can produce malodorous gas and severe esthetic problems.

The diversion of waste from one location to another with larger assimilative capacities has been practiced. The Metro project in Seattle is an example of diversion. In this case, to reverse eutrophication in Lake Washington, treated sewage effluents were intercepted and transported to Puget Sound. To satisfy the demand of the Southwest for water, it has been proposed that all the waste waters produced in the Pacific Northwest be shipped south rather than diverting the Columbia River waters southward. Technically this idea could be sound if the transportation system facilities accommodated waste assimilation during transportation. Under these conditions both regions would benefit from the diversion of waste. Probably the most recognized example of waste diversion has been developed in the Ruhr Valley in Western Germany, one of the most concentrated industrial areas in the world. About 40 percent of the total West German industrial capacity, between 75 and 90 percent of the West German production of coal, coke, iron, and steel; and some 8 million people are in the Ruhr Valley, an area roughly half the size of the Potomac River watershed. Five small rivers constitute both the water supply and the waste assimilation capacity for this area. Observations of treated waste discharges to the Ruhr River indicated that there was more waste volume being discharged than river flow. Considering the general rule of thumb that a river must have at least 8 parts of dilution water for each part of treated waste, the Ruhr was unfit for reuse, and a regional approach was taken to solve the problem. Water quality objectives for the Ruhr and Limppa Rivers were established to maintain water quality suitable for recreation and municipal-industrial water supply. The Emscher, the smallest of the three major streams, was to be used exclusively for waste dilution, degradation and transportation. Thus the Emscher was converted into a single-purpose stream, functioning essentially as an open sewer. The river channel was lined with concrete and the only quality objective was to avoid an esthetic nuisance. This was accomplished by mechanically treating effluents to remove suspended solids and then contouring the river so that it both blended with the landscape and provided aeration sufficient to avoid odors. The use of planting, gentle curving of canal streams, and attractive bridges gave the Emscher a pleasing appearance which blended gracefully into the surrounding countryside. At the mouth of the Emscher prior to its discharge into the Rhine, a large primary treatment plant was constructed for economical treatment. A secondary treatment plant is now being planned to upgrade treatment. The success of a regional plan to divert waste from two rivers to a third in order to achieve environmental quality and economies of scale in treatment has been demonstrated in the Ruhr region. It was made possible because there were three parallel streams in a very small region, and the stream being used to transport waste could be designed and regulated to avoid esthetic nuisances.

Treating the Environment

Another method of controlling environmental quality is to allow waste discharges into the environment prior to treatment of the total environment.

As demonstrated in the diversion of the Emscher River, it is possible to treat an entire river or lake as an entity rather than treating the individual waste discharge streams. This approach is used on a small scale in small lakes, rivers, canals and reservoirs. Treatment for oxygen-demanding wastes can be accomplished either by chemical addition to provide additional oxygen or mechanical reaeration or agitation to increase oxygen content. In the second case, mechanical aeration is provided by aerators—large machines which beat air into the water.

Another type of environmental treatment is the destratification of lakes and reservoirs to avoid temperature and oxygen problems. As discussed earlier, temperature-induced density gradients can cause warm layers of water to stagnate over colder layers. This phenomenon is called stratification and can be detrimental to water quality if lower waters are blocked from receiving sufficient oxygen. Intentional stratification of water is used to make the colder layers available as cooling water. Destratification provides additional oxygen and waste assimilation to lower layers, and it can be achieved either by pumping lower waters to higher levels or vice versa. In either case mechanical pumping is needed for destratification even if it is accomplished by forced aeration.

Dams provide another indirect treatment of water. Dams reduce the velocity of water flowing in rivers, and allow suspended particles and turbidity to settle out behind the dam thus clarifying the water. Dams are used to control floods, store water, and generate power, but may also cause stratification, silting and increased degradation of upstream wastes. Conflicting effects of any treatment or storage facility must always be considered when analyzing alternatives.

Treatment of land and air is not well defined. One can postulate that it would be advantageous to break up smog layers by introducing enough energy or wind to circulate the air, or to provide enough heat to dissipate fog. In either of these cases, the requisite technology is not yet currently available. Environmental control through urban renewal or zoning can effectively change the environmental quality. One can think of these as major environmental treatment processes. However, applications are too few to cite.

Desensitization

At times treatment is directed to the receptor rather than the environment or the waste stream. This practice essentially puts blinders or filters on the receptor so it is not sensitive to the environment. This is like holding one's nose to avoid bad odors, covering one's eyes to avoid unpleasing sights, or plugging one's ears to avoid loud noises. Although this head-in-the-sand philosophy cannot solve environmental quality problems, perfumes and spices have been used to disguise odors, and landscaping has hidden trash and created sound barriers to muffle noises in urban areas. While none of these solutions is attractive, densensitizing the population to polio virus with the

Salk vaccine is a positive example of this procedure. Giving hay fever shots to increase tolerance to natural air pollution (pollen, dust, etc) is another example of desensitization.

NATIONAL COST OF ENVIRONMENTAL QUALITY

Cost estimates of environmental quality protection can be very confusing. When examining these costs, several factors must be evaluated: the unit cost to reduce waste, the capital investment, the annual operating costs, and the services provided. Table 8–6 illustrates typical costs reported for pollution control. These data have appeared in government reports, industrial trade journals, the popular press, and textbooks. Depending upon the sentiments of the authors, these data tend to vary widely and may create confusion rather than resolve issues.

Examining the data in Table 8–6, one might conclude that pollution control is very costly and not feasible. However, $70 billion for 5 years would be about $14 billion per year (probably less during the first years, and more the last ones), and with a population of 200 million people, this would amount to an individual cost of $70 a year, or about 20¢ per day per person. Even if these estimates were 50 percent too low, the cost of pollution control to an individual would only be 40¢ per day. We must decide if other things that are purchased at the rate of 20¢ to 40¢ per day are more or less desirable than a clean environment.

TABLE 8–6

Cost of Pollution Control 1969-1973 (billions of dollars)

Water	
Sewer Systems	30
Municipal Treatment Plants	10
Sediment Control	6.8
Industrial Treatment Equipment	4.4
Coaling	2.1
Strip Board Treatment	0.7
Air	
Auto Pollution Control	5.9
Industrial Treatment	5.3
Research	1.6
Incinerators Control	0.3
Solid Waste	
Update Collection Systems	2.8
Closure of Open Dumps	1.2
Incinerator	0.2
	71.3

Reprinted from U.S. News & World Report, Aug. 17, 1970.

Another question that arises when examining the data in Table 8–6 is, who pays? The 20¢ per day value was obtained by assuming that each person would pay an equal share, even though some people may pollute more than others. The American consumer indirectly causes much more of the earth's pollution than citizens in an undeveloped country. Each dollar spent for a product is a vote for the pollution associated with that product. An equitable method of collecting pollution control costs is to have product prices include costs of control. In this way, consumers who purchase pollution-generating products pay the costs of control. Producers of such products claim that price increases for pollution control would provide unfair advantages for more efficient, clean industries able to market products more cheaply. Standardized nationwide environmental quality criteria have been established to prevent this, but imported products that pollute their own lands may enter our markets.

Another argument forwarded by industry is that pollution control may price the product out of the reach of most consumers. Endless studies and cost estimates of environmental quality control have failed to demonstrate that control costs are high. Table 8–7 summarizes data from government studies on the costs of clean water.

Although these data are incomplete and are for selected major waste dischargers, they indicate that price increases for pollution control should only be a few percent of the sales price. Thus current estimates of pollution control are much less than the rate of inflation of the early 1970s.

The cost of control appears to be relatively minor, but a serious problem exists regarding the rate of expenditure. In the past, neither public nor private investments have provided adequate environmental protection. As a result, many facilities have no environmental control devices. For example, one-third

TABLE 8–7

Total National Costs[1] — Water Pollution (billions of dollars)

LEVEL OF REMOVAL	10 YR. CAPITAL EXPENDITURES	20-25 YR. OPERATING COSTS	TOTAL EXPENDITURES	ANNUALIZED COSTS IN 1981
100%	94.5	220.1	316.5	21.1
80% at 95-99% 20% at 100%	47.2	110.1	157.3	12.3
95-99%	35.2	83.5	118.8	8.4
85-90% (roughly current program)	17.6	43.2	60.8	4.1

[1]Excludes $12.0 billion costs for intercepting sewers. Municipal and industrial breakdown.

Derived from Regional Construction Requirements for Water and Waste Water Facilities 1955-1967-1980, Water Industries and Engineering Services Division, Business and Defense Service Administration, 1967, in THE COST OF CLEAN WATER, U.S. Department of Interior, Federal Water Pollution Control Administration, 1968.

of the homes in urban areas have no sewers, and 7 percent of the people in these areas discharge raw waste directly to the environment. Over half the solid waste generated in urban areas is disposed of in open dumps. Many of the worst industrial polluters were built many years ago without controls, and installing controls would exceed the value of the plant. This means that a major capital investment is required just to catch up. Even though the amortized costs are small, the initial sum is large. In home construction, the costs for sewers and other waste facilities may add several thousand dollars to the price of a new home. Although the cost per day per person can be only a few cents per day, the problem of easing an initial outlay for updating older homes is serious. Major waste dischargers face a similar problem on a much larger scale. If only limited capital is available, a significant fraction must go for catch-up control. Many polluters would rather spend capital on profit making activities.

Automakers, and therefore their customers, may pay more than any other industry to control pollution caused by their products. Estimates for existing internal combustion engines indicate that $36 per control unit is spent for emission control. This will increase to $48 to provide evaporation control and finally to $200 per car to meet the 1975 standards. The $5.9 billion in Table 8–6 reflects both the costs of these controls and an equal cost estimate for use of non-leaded gasolines. Transportation pollution control may ultimately exceed 5 percent of the cost of vehicles, unless technological changes occur.

Allocation of the costs shown in Table 8–6 is only part of the study of these data. The most important question that has yet to be asked is, what is purchased for this price? In most cases the answer may not be environmental quality. The $30 billion estimate for sewers includes a large expenditure just for separation of storm drains from sewers carrying sanitary wastes. For economic reasons many cities use pipes that can handle average flows but not peak flows. During high storm flows, raw sewage overflows directly into the receiving waters through overflow outlets. Parallel pipe systems must be installed to stop these overflows. In new communities, this dual system is used initially, or storage tanks are provided to avoid overflow discharge. Although sewer costs represent almost one-half of the total control cost estimate, this action will collect only raw wastes and convey them to a few selected points for discharge. If no further investment is made, this expense will improve environmental quality in some areas and degrade it seriously in others.

The treatment costs estimated in Table 8–6 do not insure a clean environment. The water treatment costs are based on the established water quality standards, and there are many pollutants that are not specified in current standards. Mercury is not controlled, neither are toxic organic compounds, nor salts. There are standards for drinking water and for foods, but not for waste water discharges. The treatment costs are estimated to provide secondary treatment, but no more. The estimates for air quality control are only for SO_2 and particulates. Control costs for other gases and contaminants are yet to be estimated. The solid waste cost insures health protection through burial of waste, but does not provide for reuse, recovery, or environmental quality. Optimistically, since the major costs estimates are for sewer pipes, the cost for treatment of waste water, air pollutants, and solid waste should be relatively small once this investment is made.

One final remark on cost estimates. Naturally, those who must pay will overstate the costs hoping to obtain relief by delayed implementation or reduced treatment requirements. This is true for both public and private polluters. In many cases, tax relief is provided for pollution control, which stimulates the definition of all possible related facilities as pollution control. One metal refinery rebuilt its entire facility in order to support the roof top location of its pollution control equipment. Even though the facility was deteriorated and in need of replacement, the entire replacement was charged against pollution control. Some industries change their entire processes to reduce pollution and claim the entire cost as pollution control, even when the new processes are more productive and profitable. Many container industries are advertising recycling campaigns for used paper, cans, bottles, and cars. In many cases, this activity provides low cost of input material as well as some environmental control.

Although the cost to control environmental quality is hard to estimate, the damages caused by a degraded environment are almost impossible to assess. Much of the interest in control has been stimulated by threats of total disaster and extinction of man, rather than by rational cost-benefit assessments. Many health problems, for instance, can be traced to degraded environmental quality. Respiratory diseases are attributed to air pollution, water-borne diseases are linked to improperly treated or untreated sewage, and disease is carried by rodents dwelling in refuse dumps. Health costs are inaccurate and unreliable, but more direct costs, such as property damage from soiling, corrosion, and abrasion, crop damage, the cost of cleaning water supplies, and the loss of land values near dumps far exceed the estimated control costs.

In the future, complete recycling of point waste discharges may be desirable. As environmental quality standards become more stringent, or as waste discharges become more concentrated, the degree of waste treatment may reach the point where the quality of the treated effluent exceeds the quality of the air or water at the intakes. At this time, the question should arise; why discharge the effluent when it is better than the influent? Some industries are approaching this condition already—for example, those employing closed cycle cooling systems. A few towns are recycling their treated waste waters. Secondary materials industries are growing rapidly as solid wastes are reclaimed. The cost of recycling treated water, air, and solid waste is not prohibitive, and may create a new philosophy of design. Sewers are currently designed to flow downhill to treatment plants, but if the water is to be reused it may be desirable to treat it in the pipe or locate treatment facilities nearer the sources of wastes. With the large investment in sewer pipes, some thought of future use must be made now in order to avoid large costs later. The current need for sewer separation, caused by past decisions, should not continue if recycling occurs.

Even if point discharges are completely recycled, the environment will not be protected. There are numerous dispersed sources such as acid mine drainage, storm runoff, sediment and erosion from construction activities, dust and gases from fires, solid waste from logging and agriculture, and mercury and lead in rivers and lakes. In addition to physical and chemical contaminants, there is noise pollution, light pollution, electromagnetic pollution, and aesthetic pollution. These are yet to be confronted.

SUMMARY

Once a desired level of environmental quality can be measured, all necessary improvements can be defined and adjustments can be made to reduce the quantity of waste released to the environment. Adjustment may be accomplished by altering production processes to reduce waste production, treating waste streams to reduce the amount of waste released, or locating the point of discharge to minimize the impact on environmental quality. When a product itself is a pollutant, the product can often be modified, e.g., biodegradable detergents or paper. When the process creates unacceptable quantities of waste, the process can be modified so that lesser amounts or different waste are produced.

Treatment of air, waste water, or solid wastes employs four basic processes: (1) separating solids from liquids or gases, (2) separating solids from solids or fluids from fluids, (3) separating dissolved material, or (4) protecting man or ecosystems from toxic or pathogenic agents. Air cleaning has developed from chemical industry technology; waste water treatment employs natural biological processes; but solid waste treatment is basically a dump-and-burn operation. Each of these processes creates problems for the others: burning creates air pollution, air scrubbing produces water pollution, and biological treatment creates sludge that must be discarded. An integrated approach to treating all wastes is a major task for newer waste treatment systems. The problem of scale economies for large treatment facilities must be resolved by defining the proper size for treatment facilities.

The ultimate disposal of residuals from waste treatment will vary, utilizing dilution, dispersion, or detention according to local conditions. Alternatively, man or ecosystems could become desensitized to the impact of wastes.

The critical question is whether the cost to control environmental quality is balanced by the increased value of a protected environment. Current estimates seem to indicate that benefits of control are greater than costs of control.

References—Chapter 8

1. Woodson, R. D.
"Cooling Towers." *Scientific American*, May 1970, p. 70.

Suggested Readings

"The Cost of Clean Water." United States Department of the Interior, Federal Water Pollution Control Administration, 1967.

"Control Techniques for Particulate Air Pollutants." United States Department of Health, Education and Welfare, 1969.

Eckenfelder, W. W. Jr.
"Water Quality Engineering." New York: Barnes and Noble, Inc., 1970.

Mayr, O.
"The Origins of Feedback Control." *Scientific American*, October 1970.

"Waste Management and Control." National Academy of Sciences, National Research Council, Publication 1400, 1966.

"Cleaning Our Environment—The Chemical Basis for Action." American Chemical Society, 1969.

9 Policies for Environmental Control

Policy is defined as "a definite course of action adopted as expedient." The first problem in an examination of environmental control policy is to establish what is expedient. This chapter will show that a single policy that provides environmental control must contain a hierarchy of actions, each expedient for a particular environmental problem. What is currently known as environmental policy is a complex set of fragmented, conflicting policies that are as much a part of the problem as they are a part of the solution.

VIEWS OF THE NEED FOR ENVIRONMENTAL CONTROL

There is no unanimous view of the need for environmental control. Views range from "no problem" to "doomsday." Before discussing control policies, it is necessary to achieve an understanding of these views, the policies associated with each, and how these views are expressed through various professional interpretations.

Individual Views

Although there are many variations of the following views, five distinct positions exist in the current controversy over environmental quality:

No Problem. The current concern over environmental quality is part of man's efforts to survive. There is no severe danger that man will not survive; he has survived in the past, will continue to adapt, and will survive in the future. Those forecasting doom are alarmists and are wrong. *No problem exists.*

Technology Will Solve the Problems. Technology has been used by man to exploit the environment and problems do exist. Technology has not been given specific directions to correct and prevent these problems, but if so directed could easily solve all environmental problems. The problem is to direct technology toward environmental goals. *We need more science and technology.*

Back to Nature. Technology is unharnessed and running rampant. It has far exceeded the capacity of the environment to accommodate changes it has made. Cities, industries, and the society are caught in a spiral of environmental destruction. The only solution is to return to the natural state and exist within the productivity and assimilative capacity of the natural environment. *Halt technology.*

More Power to the People. Environmental problems have been caused by professionals. Laws made by professionals exploit the environment; professional technologists exploit the environment to maintain their employment; professionals are not responding to the people they serve. The people need to express their concern for the environment and regain control of the professionals. *More power to the people.*

Environment is Not the Only Problem. The environment is not a significant concern if you lack food and shelter. Clean water and air are not important if you are starving. Environmental quality is for the affluent, and is a ploy to deprive and suppress the less wealthy. Place the problems in perspective: men need jobs. *Environmental concern is for the rich.*

Some of these views may result from misinformation or ignorance, but personal values associated with facts produce these positions; therefore the holder of each of these views is inclined to adopt a policy for environmental control closely related to his perception of the problem. See Table 9–1. One of the goals of rational environmental control policy is to derive a consensus of these views and point out problems recognized by all. For example, if existing policy is regarded by many to be ineffective, then a new policy may be needed. The purpose of this chapter is to present these policies and determine their performance for each of the different views of environmental problems.

Professional Interpretations

Value-related policies such as laissez-faire, more science, back to nature, power to the people, or jobs first represent only one type of policy encountered in environmental quality control. An alternative policy classification could focus on disciplinary definitions created by the professions that administer environmental policies. Whereas many discipline policies are not unique to environmental issues, they present particular characteristics that are significant in resolving environmental issues. Briefly, these classifications are:

Legal. Innocent until proven guilty is a policy that is fundamental to our legal system. Expertise is required to identify those who are at fault, and until fault can be proven no remedy can be obtained. In cases of irreversible damage, such as the extinction of a species or destruction of an ecological community, the policy of proof before action will not protect the environment.

Political. The current political process has been described by Lindbloom and Braybrooke (see list of suggested readings) as muddling. These authors characterize the political process as one which is concerned with minor changes made with limited information or none at all. Muddling depends upon the attentive presence of many observers and critics who voice objections when the change is not desirable. Any change should be small enough so that if it is in the wrong direction, it can be corrected without serious costs. Muddling is endorsed by environmentalists, since major catastrophes can be avoided. However, progress may be slow.

Planning. Planning attempts to foresee problems and avoid them. Such action requires information and administrative power that may be impossible to acquire. An alternative role for planning is to provide information for political or legal processes and to employ non-planning policy to achieve planning goals. Planning is associated with the scientific-technical community and has not in the past been an effective force for environmental control.

Economic. The market place is an effective force for allocating the resources of this country, but it has failed in many environmental resource management problems. Economists such as Kneese (list of suggested readings) have identified remedies for economic policy to provide environmental control. Proper project evaluation of environmental considerations may be possible in economic terms, but careful analysis is required.

Administrative. Government agencies charged with protecting the environment may view with frustration the failures, trials, and errors of legal, political, and economic policies. Rather than relying on methods that may cause environmental damage, anti-degradation methods or conservation may be employed. This "better-safe-than-sorry" approach will retard economic growth and be attacked by those valuing profit and growth over environmental quality.

This brief description of two classifications of environmental policy will now be expanded with examples to search for common issues that must be faced by everyone. A summary of these policy categories is provided in Table 9–1.

TABLE 9–1

Policy Matrix—Search for Consensus

VIEW OF ENVIRONMENTAL PROBLEM	VALUE RELATED POLICY	DISCIPLINE RELATED POLICY: LEGAL	POLITICAL	PLANNING—TECHNICAL	ECONOMIC	ADMINISTRATIVE
No Problem	Laissez-faire					
Technology Can Fix	Increased R & D					
Back to Nature—Doomsday	Stop Technology—Need Diversity					
More Power to the People	Grass Root Action					
This is Not the Only Problem	Jobs First—Environmental Quality Later					

TEST ISSUES

Three hypothetical issues are presented here in order to provide a basis for analyzing various policies. The first issue, labeled "*Skunk Works,*" represents large single pollution sources that discharge uncontrolled waste. Characteristics of interest for this issue are presented in Figure 9–1. The second issue, labeled "*Rapid Growth,*" represents the sprawl and uncontrolled growth of many urban areas. Figure 9–2 summarizes useful data. Rather than introducing the complexity of an actual situation, we use the simple problem of filling a volume with waste from increasing sources. This illustrates many problems without confusing issues with technical concepts.

The third issue, characterized in Figure 9–3, is called "*Blacktop,*" and represents the environmental changes caused by major construction or public works projects. Again a simplified situation is used to present policy problems of major works such as the SST, the Alaska pipeline, and the Everglades Airport. No positive scientific data are available, and all information is based on experience and judgment from existing data about similar projects.

APPLICATION OF INDIVIDUAL VIEWS TO TEST ISSUES

No-Problem View

The first analysis examines the no-problem policy to determine if the discipline policies have anything in common for each of the test issues. The

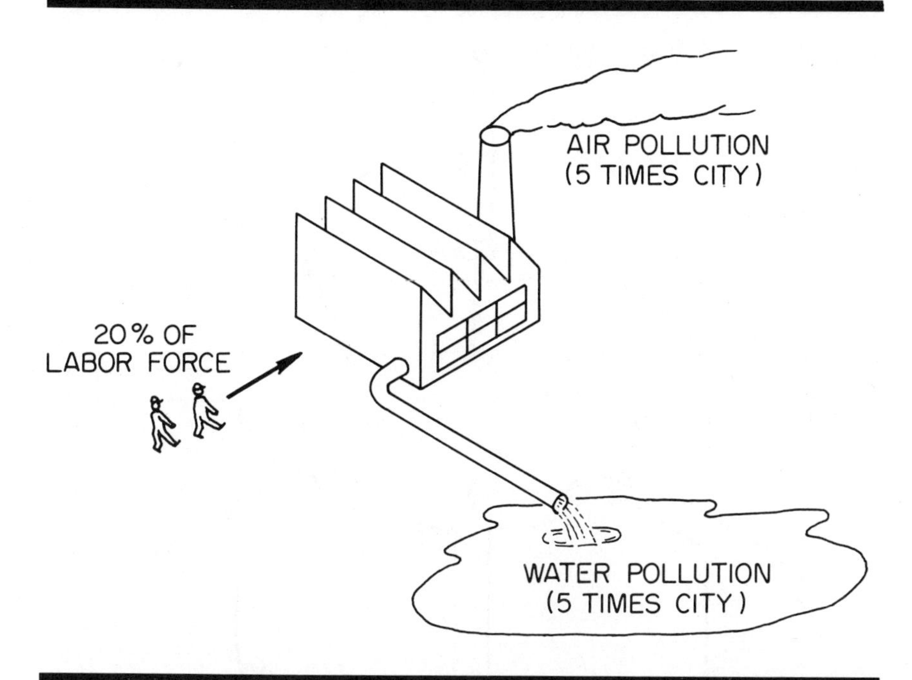

Figure 9–1. Characteristics of Test Issue No. 1—Skunk Works
Plant is 20 years old.
Plant produces water and air pollution that is five times that of the surrounding community.
Management plans to double capacity of plant.
Cost of reducing waste 90 percent from old plant equals value of old plant.
Cost of reducing waste 90 percent from proposed plant is 5 percent of total cost of new plant.
Plant employs 20 percent of work force in the community.
Legal costs are 1 percent of plant value per year to avoid clean up.
Not all discharges are covered in standards.
Environmental quality is below standards now.

no-problem view contends that man is hardy and resourceful and that he can endure environmental change and avoid disasters. Under this policy any changes or reactions are considered part of man's natural ability to cope with problems. More conservative subscribers to the no-problem policy would be against any changes, since the current way must be the best way. In all three test cases, this policy would result in the use of existing procedures to effect decisions. This policy still requires an information system that can identify and solve problems in the existing framework.

The *Skunk Works* will not be considered a problem. The plant management has proposed a plan to reduce wastes and only asks a 5-year delay to insure that the technology can be tested prior to implementation. *Rapid Growth* has not reached a problem status and very few are concerned at this time. Environmental protection forces are currently sounding the alarm and some-

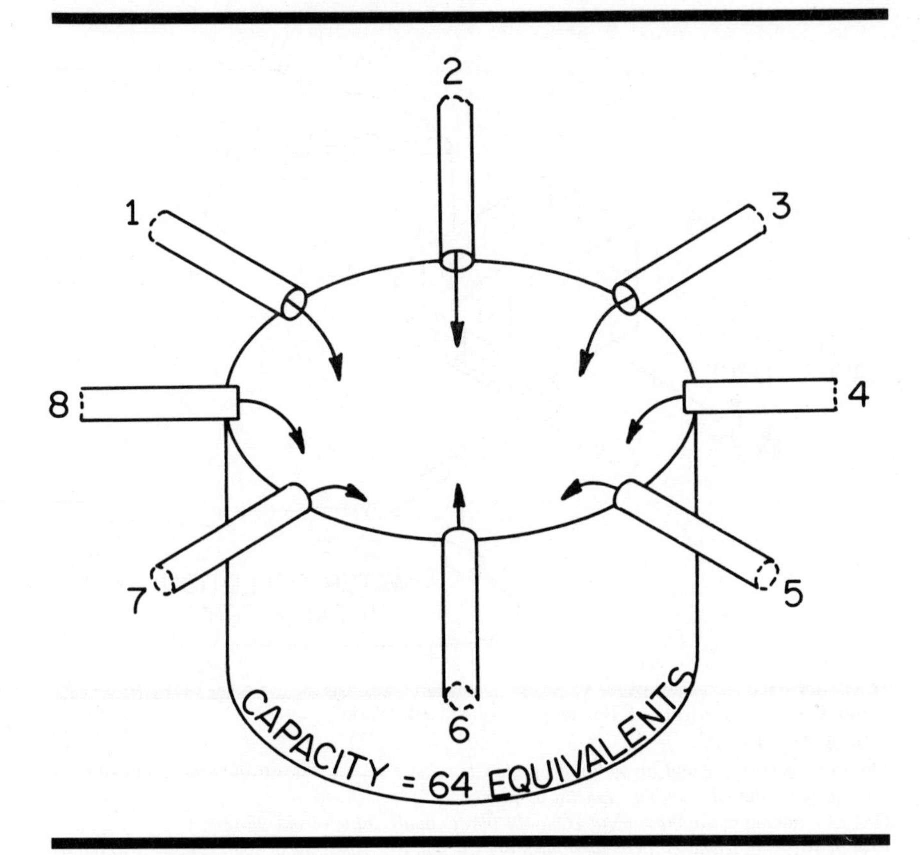

Figure 9–2. Characteristics of Test Issue No. 2—Rapid Growth
A regional assessment of waste assimilation indicates that 64 waste equivalents are the maximum that can be released. Currently there are eight dischargers, each releasing one waste equivalent (8 total). Waste production has been programmed to increase:

Time from now	*Waste equivalents/ unit time/discharger*
0	1
10	2
20	4
30	8
40	16

Cost of waste abatement for 50 percent reduction:

Total Waste Production	*$/Waste Equivalent*	
Waste Equivalent	*Individually*	*If all combine waste*
8	1×10^6	2×10^6
16	1×10^6	1.5×10^6
32	1.5×10^6	1.5×10^6
64	2×10^6	1.2×10^6
128	3×10^6	1×10^6

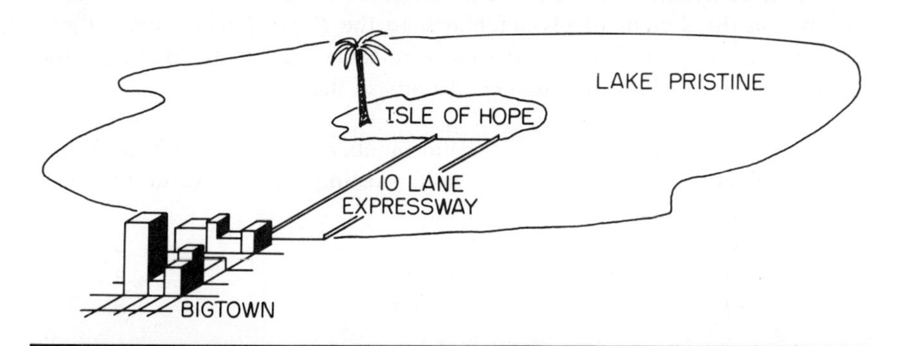

Figure 9–3. Characteristics of Test Issue No. 3—Blacktop

Proposed to pave Isle of Hope for airport for SSTs and build ten-lane expressway to connect to Bigtown at federal government expense.

Science Board A testifies:

1. *Island is last nesting ground of Green Belly Weight Watcher.*
2. *Expressway will stop circulation in lake and cause eutrophication.*
3. *Noise will cause all wildlife to leave for twenty-mile radius.*
4. *Exhaust from cars and planes will increase lead in lake.*

Science Board B testifies:

Board A is alarmist and project will enhance recreation potential since island was unproductive swamp.

thing will be done when the time comes. The *Blacktop* issue presents no problem since both sides are active and administrative forces will provide a satisfactory compromise.

Technical Solution Policy

The policy relying on technology can be employed by both sides in an environmental confrontation. Those supporting the proposed project will argue that much more technology is required to define any potential degradation and specify solutions. Those advocating that the activity could damage the environment will argue that the technology is available and can be applied. In the case of the *Skunk Works* issue the environmentalist will demand research and development to identify the discharges from the plant, the effects of these discharges on the environment, the development and demonstration of equipment to avoid such discharges, and a detailed system of monitoring. Those associated with the expansion of the *Skunk Works* will advocate that the existing standards define the environmental protection required and that they will employ "all known, available and reasonable methods of treatment." They will contend that if and when the government can demonstrate the need for control of other contaminants in their waste, they would provide the additional treatment.

The *Rapid Growth* issue presents several problems that are not involved in the *Skunk Works* issue. One problem is that the amount of waste discharge to be treated depends not only on the actions of a given waste discharger, but also on the actions of his neighbors. In the *Rapid Growth* issue, there is a total capacity of 64 units. At time zero there is 8 units and every time multiple of 10, doubles the number of units so that:

Time	Total number of units existing	Waste from waste producer
0	8	1
10	16	2
20	32	4
30	64	8

A single waste producer will view the total available capacity of 64 units and his projected waste discharge of 8 units at the end of 30 time units. If there is no monitoring of dischargers or if an individual discharger is unaware of others' actions, one discharger would assume that no problem exists until his waste approached the limit for the region.

Even if waste discharge can be observed and the assimilative capacity of an area is known, some policy must be evolved that will define "who dumps what, where, and when" and "who pays for finding out who is dumping what, where, and when." Technology can provide alternatives for dumping, but science and technology have not addressed the problem of allocation. In the *Rapid Growth* issue, technology could develop methods to observe increases in waste discharges, to delay these increases by providing increased abatement technology, and to provide substitutes for products that create waste. The technology policy cannot provide environmental quality with a policy on how to use technology.

Blacktop presents additional but similar problems for the technology policy. The problems of evaluating technical alternatives, and of the inability of technology to predict accurately the outcome of its products are emphasized by *Blacktop*. There are many environmental issues that have not been studied in the past and may require many years of research to resolve. In such cases, technologists can venture their opinion of the possible outcome, but actual outcomes must await the test of time. In the case of the supersonic transport, many of the environmental problems cannot be identified. "Try before you buy" is the technology equivalent of the "innocent until proven guilty" policy of the legal profession. Technologists assume that for any problem, given enough time and enough money, a solution can be developed. This policy may reaffirm the no-problem policy.

Back-to-Nature Policy

Many concerned with environmental quality doubt the compatibility of technology and ecology. This view can be briefly summarized as "technology is waging a losing battle—every new development creates another series of environmental problems" or "in the long run man cannot improve on nature." Actions which provide short-term improvements on nature seem to result in long-term environmental problems that are more severe than the original ones that were solved.

The *Skunk Works* issue is viewed in this context as a prime example of environmental exploitation. In the rush for jobs and profits, the community has foregone environmental quality and long-term impacts are appearing. Each time that plant management presents a concept for controlling waste discharge, control is provided at the expense of another part of the environment. In the case of the *Skunk Works*, the air pollution control proposal resulted in a water pollution problem and vice versa. Simultaneous control of air and water pollution created a solid waste problem.

The back-to-nature forces view the *Rapid Growth* issue as typical of uncontrolled growth. Each individual discharger may be challenged by these advocates but they usually lack resources. In the long run the back-to-nature forces believe that technology will prevail, the environment will be ruined, and man will become extinct. The only hope is to stop technology; no other action can provide control. The *Blacktop* issue only confirms these views, they believe, for even the technologists cannot agree in this case on the outcome of these projects. The back-to-nature forces do not want to risk environmental damage in favor of more development, more growth, and more affluence.

More Power to the People

The case for this policy cannot be adequately understood until the policies of the professionals are understood. In the following sections, the professional policies will be examined in detail and references will be made for "people power" in those discussions.

Environment is Not the Only Problem

Before examining professional policies, the final control policy is presented. The *Skunk Works* issue employs 20 percent of the work force in the community. If the plant is doubled, there will be more jobs and more people attracted to the community. If the community is isolated and has a large fraction of the work force unemployed, the citizens of the community may view the project as desirable. There is a chance for jobs for their families; their children will not have to leave town to find work; the tax base will increase to provide new services; family incomes will increase, and so on. Although environmental quality may suffer, the people of the community may perceive a net economic increase.

The *Rapid Growth* issue presents a larger view of the *Skunk Works* issue. In this case, growth may not only create environmental quality problems, but the influx of new jobs and higher affluence creates social problems such as crime, imbalance in education and housing, minority problems, etc., which are created by the larger population. Minority groups can then reach the critical mass where they can make a major impact on policy issues. Even though environmental quality is of concern, it is not as important as other urban problems. The *Blacktop* issue could present the opportunity for the community of Big Town to become the gateway to the world at no monetary expense to the region. Federal monies are available for the project through the efforts of the region's congressional delegation. The region can have a new economic base, local leaders can acquire new political stature, the business community will profit. Can environmental concern be significant compared to all the benefits from the project? Many would argue that the environment can be protected and permit the project, others would say the benefits are less than the costs and reject the project.

POLICIES OF THE PROFESSIONALS

Members of each profession will hold a wide range of views in philosophy. Gross generalizations are made in this discussion to emphasize the pros and cons of these opinions. We will not examine the views of theorist versus practitioner or conservative versus liberal within each group.

Legal. The *Skunk Works* issue can be used to illustrate several problems with existing legal policy, if environmental protection is a goal. Given the policy of "innocent until proven guilty," the *Skunk Works* can increase its discharges until data are obtained proving that the discharges do in fact cause damages. If standards have not been established, then additional time may elapse before legal action can be taken. Even when the case finally reaches the courts, the damages caused by the discharges will be weighed versus the potential loss to the community if the *Skunk Works* were to close. This type of threat has been common in single industry towns, and the legal system makes little provision to compensate a community when the major employer leaves as a result of environmental protection. In the case of the *Skunk Works*, the cost of legal action to delay investment is much less than the cost of abatement. This fact provides an additional incentive to delay and permit potential environmental degradation. The second issue indicates the significance of time in legal resolution of environmental issues. If legal action is taken against each discharger in turn, and the time for resolution exceeds 3 time units each (24 units total), the total waste discharge will exceed assimilative capacity before the issue of abatement can be resolved. A legal problem that may occur with *Rapid Growth* concerns the development of legislation to permit regional rather than individual processing of waste. Cost data indicate that economies of scale exist in waste treatment if jurisdictional problems can be resolved. The *Blacktop* issue presents the problem of preventing damage before it occurs. The ability of legal policy to resolve environmental problems depends on political and administrative policies to establish new laws. Change of environmental law from "innocent until proven guilty" to "guilty until proven innocent" may be necessary to preserve environmental quality.

Under the existing court system a consistent policy cannot be developed. Each case is heard by a different judge and jury attempting to employ a given set of rules to special cases. This does not lead to a consistent policy or to a long range plan.

Political. Before discussing the three issues from a policy viewpoint, the classification of strategies presented by Braybrooke and Lindbloom may provide insight. Figure 9–4 represents their grouping of issues in terms of the quality of understanding and the size of change resulting from implementation of a decision. The first quadrant contains issues that are well understood and result in large changes. "Utopian decision-making" is the proper title for such issues since few if any environmental issues can qualify. The second quadrant of Figure 9–4 contains issues that are well understood and cause incremental changes. Such issues can be resolved by technology. The third quadrant contains the bulk of environmental issues: the understanding is

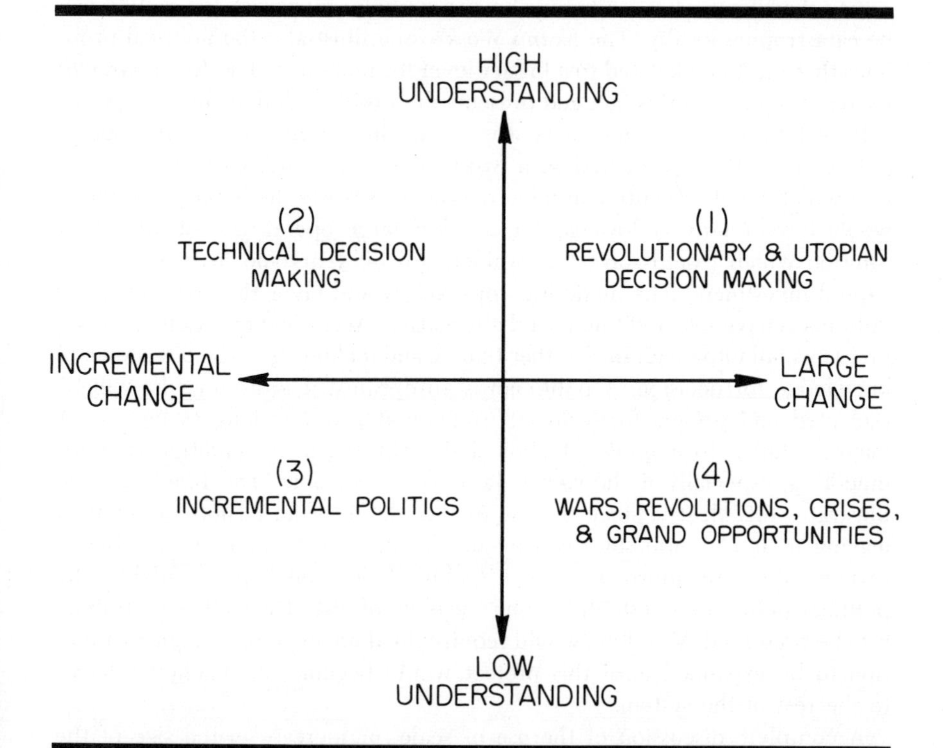

Figure 9–4. Classification of Issues

From A STRATEGY OF DECISION—POLICY EVALUATION AS A SOCIAL PROCESS, by D. Braybrooke and C. E. Lindbloom. Copyright © 1963 by The Free Press of Glencoe, second printing, p. 78.

incomplete and changes are incremental. The final quadrant is self explanatory. These issues should be avoided by preventive actions and not allowed to reach confrontation proportions. Although it is the goal of a technologically oriented society to move issues into the first or second quadrant (where rational and informed decisions are possible), many issues in the third quadrant must be resolved immediately. As pointed out earlier in this chapter, Braybrooke and Lindbloom contend that the most effective policy is to "muddle."

The *Skunk Works* issue presents some challenging problems for the political analyst. The technologist can provide abatement alternatives, but the politician must decide "how clean is clean" and he must determine reasonable methods of abatement. If the politician has jurisdiction only over the local area, he faces the potential of a large change rather than an incremental change. If he requires too much abatement, the *Skunk Works* may move to another region. This would result in a major loss of employment in the immediate region. On the other hand if the *Skunk Works* were allowed to grow, the environmental quality might degrade so that other industries would not enter the region. The politician must have the ability to regulate small rather than large changes before he can muddle, otherwise the results can be catastrophic locally. The *Skunk Works* issue illustrates the political problems that must be elevated to a higher level for muddling. The *Rapid Growth* issue re-emphasizes this fact and presents the problem of dynamics for political resolution. If the time units were years in the *Rapid Growth* issue, a politician could be presented with two technological alternatives: (1) a solution which would require continual decisions as waste discharges and which would have relatively low capital cost but large operating cost, and (2) a solution which would solve the problem for 50 years and involve a major capital investment. The muddling approach would favor the first, but many politicians have selected the second alternative. According to rationale, waste problems fall into quadrant 2 rather than 3, and technology provides adequate knowledge for decision. A politician is an optimist who assumes he will be reelected and prefers to avoid environmental issues as long as he can. A major pitfall is to employ a technical strategy when the situation calls for muddling, especially if the step is large or irreversible. The *Blacktop* issue illustrates this clearly, since the use of the island for an airport rather than leaving it in a natural state is a major change that is almost impossible to reverse. Since all information for a technical decision is not available, the political policy must establish some measure of risk if a technical strategy is to be preserved. Muddling would require the domain of impact and jurisdiction to be expanded until this project would become incremental relative to the rest of the system.

A complete discussion of the use of scale, or increase in the size of the system, to achieve a basis for muddling will be presented in the evaluation of administrative policy. At this point, it should be recognized that the upper limit for such a strategy would be the earth itself. When the system for control is the earth, the problem takes the form of the "spaceship earth" described by Boulding in *The Environmental Handbook* (see suggested readings).

Planning–Technical. Planning and technology base their policy on the ability to elevate all issues to the second quadrant of Figure 9–4. At the end of the 1960s science was a dominant force in our society and many decisions that had major impact on society were made by a technical policy—the development of data and the rational selection of optimum choices based on quantitative data. The planning movement cited the confusion, redundancy, and inefficiency resulting from previous political muddling and advocated "planning ahead." The earlier discussion of the technical solution policy is also applicable to this discussion.

If a planning-technical policy is to operate successfully, several characteristics must be present that do not exist today. The first characteristic is the ability to forecast the outcome of any action, or at least the ability to assign probabilities of alternative outcomes. This in turn requires that quantitative assessment be used to evaluate alternatives. Another characteristic which is now neglected is the ability to examine the result of a decision and take corrective action. If the technical approach were used in any of the three test issues and a decision were made, it would be important to have a system to monitor the performance of the decision. For example, if the airport in the *Blacktop* issue were to be constructed, some system of environmental surveillance would have to be available to determine if environmental damage occurred. It is one thing to predict that no damage will occur, but it is another to ensure that it does not.

The most important point in using technical strategy is to recognize the limits of quantitative assessment; moreover, subjective values must enter at some point in the analysis. In the *Blacktop* issue, the more of the analysis that is quantitative, the less confrontation there will be in resolving subjective values. If it can be proved that quantitative ecological changes will result from the airport, then the subjective evaluation can be made between the changes in ecology and the benefits of the project. On the other hand, if the changes are only suspected, both the probability of change and the value of change must be assessed.

The three test issues do not reveal the major fault of the technical approach. In the 1960s, the technical forces became large monopolistic powers that could propose and obtain approval for projects that created large environmental changes. These powers were to develop single-purpose projects with little concern for the impact of one project on other needs of society. There were highway builders, dam builders, navigation and port developers, resources developers (fisheries, forestry, oil, coal, etc.), and the military establishment. Each of these technological powers employed the technical policy, and each has failed to consider the social impact of their efforts.

As a result, in the 1970s science has fallen into disfavor with society for its insensitivity to human needs.

Economics. The economic discipline has contributed much to the downfall of the technical policy and has provided insight as to why it has failed. The technical policy is a product of the capitalistic market system studied by economists. One economic policy uses the market place to allocate uses of the environment. Students of economics who apply the theory of the market place to environmental problems soon conclude that the many assumptions necessary to allow the market place to operate correctly cannot be satisfied in environmental allocation problems. In the *Skunk Works* issue, if the plant discharges wastes that degrade the environment of the community, there is no market mechanism which charges the *Skunk Works* for the use of the environment. Costs that are not borne by a firm in its production activity are defined as externalities. Naturally anything that is free will be used much more than an equivalent thing which is costly. In the case of a natural resource, whether it be water, air, fish, or oil, it will be used or abused if it is free but conserved or used judiciously if there is a charge for use. No policy advocated by the economist will place a proper charge for the use of the environment. In the case of *Rapid Growth*, if a charge for waste disposal were greater than $\$3 \times 10^6$ per unit per time period, no one could afford to discharge without first treating the waste; it would not pay.

A similar policy has been suggested in the evaluation of projects. The benefit-cost analysis was developed by the technical-economic professions to provide quantitative assessment of projects. Unfortunately, benefit-cost analysis can be abused by claiming many benefits that are not realized or by double-counting benefits, by not claiming all costs and realizing externalities, and by employing interest rates that make the discounted project values appear favorable. A complete discussion of this can be found in the readings listed at the end of this chapter.

Another problem identified by the economist is the "common pool" characteristic of the environment. The market place requires goods to be allocated so that each buyer can purchase as much as he wants. In the case of water or air in a given region, this is not possible. In the *Skunk Works* issue, the firm cannot buy dirty water and, at the same time, the town cannot have clean water. The one good (water) must serve two parties (the firm and the community). Economic policy suggests several alternative actions which require institutional changes to permit the market to operate. The first alternative, discussed earlier, is to have a third party own the water and charge for its use. The second is to reorganize ownership so the firm and the community operate jointly for the profit of both (this is called internalizing an externality). Another solution is to define a limited number of rights to discharge that can be purchased. Then if clean water is desired the rights can be purchased and not used, thus reducing waste discharge.

Another reason for cooperation is the ability to reduce unit cost by combining activities. In the *Rapid Growth* issue, it is cheaper for all waste dischargers to combine their waste in a single plant than for each to provide treatment individually. This cooperation is more common for water than air dischargers, but even with air there is a tendency to build one large production facility rather than many small ones. Economic policy would suggest the need to reduce roadblocks that prevent consolidation to realize cost savings. The roadblocks are real and substantial in many environmental problems and require a major effort for removal.

Administrative. Given a charge to maintain a clean environment, policy must be established to achieve this goal. This policy must consider implementation, regulation, and refinement aspects. One administrative policy could establish a uniform policy for all dischargers. In the *Rapid Growth* issue, the administrative policy could establish a uniform standard for nationwide environmental quality (this would prevent a discharger from threatening that he will move to another locality with less stringent standards). Another implementation policy could require all dischargers to provide the same level of treatment. The rationale behind this argument is one of equity since all dischargers must take the same action. The fault with this theory, of course, is that each discharger has different capital investments and technology bases as well as different locations. One plant could provide the required treatment at much lower costs than another, or the assimilative capacity at one location could be much larger than at another. Thus, in practice, uniform treatment does not ensure equity. Another disadvantage is that uniform treatment discourages alternative abatement methods such as process change, product modification, etc.

In the enforcement of policy, the administration must decide whether to monitor the dischargers. A self-monitoring policy is often used. Each discharger is required to measure and report discharges periodically. This is about as effective as requiring motorists to observe and report monthly the number of times they exceed the speed limit. The major problem with monitoring is the lack of funds. The self-monitoring policy is usually a result of limited funds. An administrative policy that attempts to regulate environmental quality must have an aggressive monitoring and enforcement program.

If feedback is not available for control, no conrol is possible. The back-to-nature policy assumes that nature's control using diversity and small perturbations is much better than the large fluctuations that can occur when human efforts fail. From the administrative viewpoint, a policy for environmental control is moderated by a policy of agency survival and operation. All policies may be incorporated into the administrative policy. Stress may result in compromises which provide political acceptance but allow continual environmental degradation.

One issue that must be faced in administrative policy is the size of the jurisdiction area. The ability to internalize problems by increasing regional authority must be balanced by the losses in communication in larger organizations. The power-to-the-people movement is in direct conflict with economies of scale policies. If the technical aspects of these problems can be resolved, the human values still remain. The administrator, faced with such problems should seek a policy that can protect the environment as well as maintain his employment.

POLICY RESOLUTION

This chapter has attempted to present various policy positions that exist in environmental management. The use of hypothetical issues to illustrate these policies does not mean that the policies are hypothetical. Real case histories can be identified that have all the characteristics of these test issues. Tables 9–2 and 9–3 are summaries of the various policies analyzed. Fundamental to each policy are the particular assumptions concerning the available information, the ability to forecast the future, and the ability to avoid unforecasted events. The policy views are to muddle or to be scientific. Conflicts appear when one attempts to muddle scientifically or to indulge in scientific muddling. In either case, the expectations of the policies are not satisfied. When muddling is used there is a basic demand for feedback and for control to enable reversal of an action. Neither of these conditions is met by present environmental policy. On the other hand, full information is required when a technical policy is used. The inability of current technology to identify and measure the values of all policy impacts has contributed to the lack of confidence in technical solutions. Not all impacts can be quantified at this time, and policies based on assumptions that eliminate the need for such quantification will fail. The assumption that only direct costs of environmental quality change are significant will not prevent indirect costs from occurring.

In the previous discussion of policies, little consideration has been made of criteria and goals for environmental policy. This omission is not an oversight, but a void in management. It is assumed that environmental quality is guaranteed when a standard or treatment level is established. An examination of the historical shifts in environmental quality will prove the naiveté of such assumptions. Environmental policy must be evaluated in terms of performance criteria, such as: the divergence of quality from standards induced by the policy, the response time required by the policy to correct degradation, and the costs or benefits resulting from the use of a given policy. No examples of policy performance using these quantitative criteria have been reported. What has been reported is the inability of past policies to provide environmental quality.

In summary, environmental policy has been developed through muddling. Each unsuccessful policy has led to an alternative policy. There exist many competing demands for resources available to formulate and implement envi-

TABLE 9–2

Summary of Test Issues — Individual Views

	ISSUE		
VIEW	SKUNK WORKS	RAPID GROWTH	BLACKTOP
No Problem	No problem; system will decide	No problem; system will adjust	No problem
Technical Solution	■ Need R & D ■ Solution available ■ Problems only need to be defined	■ Need feedback ■ Need user of technology	■ R & D takes time ■ Who values outcome ■ May be some risk
Back to Nature	One solution creates another problem	Growth too fast to need reduction in technology	■ Technology uncertain ■ Gone too far already ■ Need to manage technology
People Power	Local decision needed	Individual decision vs. regional decision	Federal interests, not those of community
Environment is not the only problem	Jobs first	Crime, housing, conflict more important to resolve	Government aid cannot be rejected; growth is good

TABLE 9–3

Summary of Test Issues — Professional Views

VIEW	ISSUE		
	SKUNK WORKS	RAPID GROWTH	BLACKTOP
Legal	■ Prove damage, new standards ■ Balance benefits vs. economic loss ■ Legal cost small	■ Time to halt ■ Courts do not plan ■ Response time	■ Cost of action high ■ Lack of precedence
Political	■ How clean is clean ■ Jobs first ■ Equity	■ Regional approach vs. local ■ Short term vs. long term fix ■ Muddling accelerated	■ Large charge ■ Need for feedback ■ Irreversibilities ■ International implications
Planning—Technical	■ Land use ■ Environmental Models ■ Control & feedback	■ Information systems ■ Real time monitors ■ Rapid response	■ Technical forecast ■ Avoid problems—scarcity ■ Social management of technology
Economic	■ Externalities ■ B/C—technology assessment ■ Public goods	■ Economies of scale ■ Resource allocation	■ Social costs ■ Interest and discount
Administrative	■ Anti-degradation ■ Uniform standards	■ Hit or miss enforcement ■ Monitoring	■ Organization and responsibility ■ Multiple objectives

ronmental policy. More understanding of policy issues is required if a rational technical approach to policy is desired, but there is no guarantee that such an approach will be more or less effective than existing methods. Criteria should be established to evaluate the existing policies. Without measures of performance, policy evaluation is academic.

It would be a useful exercise for the student to trace the historical pattern of local and state environmental management agencies and to define what policy issues predominated in each decision to reorganize management institutions. In many cases, the states will take the dominant role in implementing standards, whereas federal agencies will focus on the national uniformity of standards and the development of a technological base for state application. Local environmental functions are also shifting to the state level because of the necessity for regional rather than local control of land use and waste discharges.

At the federal level, the management of environmental quality has shifted from public health protection in the first half of this century through control of drinking water standards, vectors (solid wastes), and water-borne diseases, to the control of water and air quality, the management of environmental quality, and finally to the concept of a "contained spaceship earth" that has only limited resources. A brief summary of Federal Legislation and directions is presented in Table 9–4.

The principal federal organizations of the 1970s are the Environmental Protection Agency and the Council of Environmental Quality. The Environmental Protection Agency (EPA) was created from agencies that were concerned with water quality (Federal Water Pollution Control Administration), and from air and solid waste agencies that were originally in the Public Health Service. The FWPCA was itself a spin-off from the Public Health Service earlier in the 1960s. In 1971, the EPA was made responsible for controlling noise, increasing its areas of jurisdiction from five (air, water, solid wastes, pesticides, and radiation) to six. The EPA has the responsibility for setting and enforcing standards, for developing technology to abate pollution, and for satisfying the standards through research, development and demonstration projects.

The Council on Environmental Quality is part of the Executive Branch Office and has the primary role of advising and assisting the President on environmental policy. The National Environmental Policy Act of 1969 requires all federal agencies to take full account of environmental factors in their planning and decision making. As part of this responsibility, each agency must develop and file with the Council an analysis for every proposed action that could have significant environmental impact and provide these statements to federal and state agencies and the public. The Council does not have authority to stop or alter a program, but it can advise the President.

One of the most important bases for environmental management is the Rivers and Harbor Act of 1899 which requires anyone discharging waste

TABLE 9–4
Brief Summary of Federal Water Quality Legislation and Directives

Year	Legislation
1899	Rivers and Harbors Act of 1899 (protect navigable waterways from obstruction)
1912	Public Health Service Act of 1912 (authority for PHS to conduct pollution investigation)
1924	Oil Pollution Act of 1924 (prohibits oil discharge in coastal waters)
1927	Rivers and Harbors Act of 1927 (authorized US Army Corps of Engineers to make river basin studies)
1946	Conservation of Wildlife Act of 1946 (protect wildlife resources)
1948	Water Pollution Control Act of 1948 (P.L. 485) (expand PHS activities)
1956	Federal Water Pollution Control Act of 1956 (shifts responsibility to states)
1958	Water Supply Act of 1958 (shifts responsibility to states)
1962	President's Water Resource Council formed 1962 (develop uniform policy for Federal Departments)
1964	Water Resources Research Bill, 1964 (created OWRR to sponsor research)
1965	Water Resources Planning Act (establishes Water Resources Council & River Basin Committee)
	Water Quality Act of 1965 (creates FWPCA in Department of Interior, states must set standards)
1966	Clean Water Restoration Act of 1966 (Revise Oil Pollution Act of 1924, provides grant monies)
1970	Water Quality Improvement Act of 1970 (control of oil and hazardous wastes)
	EPA established
	National Environmental Policy Act of 1970 (established CEQ and requires environmental impact statements)

to obtain permission from the Secretary of the Army. Unused until the late 1960s, it provided a legal tool for environmental management of waste discharges. The Resource Recovery Act (1969) reflects a concern that management of environmental quality must also consider conservation of resources. Waste discharge not only degrades the environment but prevents recovery of resources that are becoming scarce.

The progression of environmental management from public health to resource management is reflected by organizational changes. In the early 1970s about 30 percent of the states formed organizations charged with the management of environmental quality. As many as 60 percent of the states still have environmental management fragmented into agencies concerned only with water or air or solid waste, and many of these agencies are still in health departments. There are a few states that have combined environmental quality with natural resource management, and one, Michigan, has given every public or private entity the right to sue any other entity in the state courts to protect the environment. The mechanism of class action lawsuit by private citizens is an additional institutional approach to environmental management.

To understand the process of muddling toward environmental control, the student should examine the progression of environmental management in his area to identify the focus of each agency, the type of control, and the new problems created by each new institution.

SUMMARY

This chapter has presented a series of hypothetical issues to illustrate the conflicts that arise between various sectors of our society and the bureaucratic solutions proposed to resolve these conflicts. The issues represent the problem of the one industry town with serious environmental degradation, the environmental degradation associated with continued growth, and the problem of major federal or multi-industry projects such as the SST. The various views examined are: (1) no problem, (2) technology will solve it, (3) back to nature, (4) power to the people, and (5) the environment is not the most important problem of our times. Interwoven with these views are discipline-oriented solutions offered by the lawyer, politician, planner, economist, and administrator.

The misinformation and misunderstanding that create these views are explored with the test issues, and it is concluded that this diversity has produced a muddling approach to environmental control. In the future, unless goals are better defined and more information can be obtained more rapidly, this type of control will ultimately fail. Unprecedented disaster could result.

Suggested Readings

Braybrook, D. and Lindbloom, C. E.
A Strategy of Decision—Policy Evaluation as a Social Process. New York: The Free Press, 1963.

Dales, J. H.
"Land, Water, and Ownership." *Canadian Journal of Economics*, November 1968, pp. 791–804.

DeBell, G.
"The Environmental Handbook." New York: Ballantine Books, 1970.

Goldman, M. I.
Controlling Pollution—The Economics of a Cleaner America. Englewood Cliffs, N.J.: Prentice-Hall, 1967.

Herfindahl, O. C., and Kneese, A. V.
Quality of the Environment; An Economic Approach to Some Problems in Using Land, Water, and Air. Resources for the Future, Inc., 1969.

Kneese, A. V.
The Economics of Regional Water Quality Management. Baltimore: The Johns Hopkins Press, 1964.

McKean, R.
"Cost Benefit Analysis and Efficiency in Government." Santa Monica, Calif.: Rand Corporation, 1955.

10
Epilogue

ARGUMENTS FOR POLLUTION

The authors have sometimes had the privilege of visiting primary and secondary schools in the Seattle area to discuss pollution and to answer questions of schoolchildren. On one occasion, after a 15 minute presentation of the possible bad effects of increased heat and carbon dioxide in the earth's atmosphere, a fifth grader approached an audience microphone and politely asked, "Mister, if we are polluting the air so badly that we may be hurt by it, *why are we polluting the air?*"

Let's assume for the moment that there is a reasonable answer to this question. The reasons for pollution might include:

Apathy–Ignorance. Pollution exists because too few people appreciate the magnitude of the short and long term pollution threats and therefore exert no social or political pressure to alleviate them. The solution would be a challenge to communication and education.

Inadequate Technology. Pollution exists because we simply do not currently possess the scientific and engineering know-how to stop it. The solution would require more investment in technological research.

Legal Right to Pollute. Pollution exists because use of natural resources is not legally controllable. A pulp mill has every bit as much right to use the water of Lake Pristine as you or we do to row a boat on it. The solution would require a challenge to the legal-judicial system.

Better Uses of Resources Exist. Pollution exists because our air, water, and land resources must be used for many purposes, including the disposal of wastes. An excellent example of this argument became apparent during the

Apollo moon landings when, to the horror of some, millions of dollars worth of photographic and other equipment was left on the moon at the end of the mission. The reason—it was cheaper to leave it than to build a return vehicle large enough to bring it all back. The solution would require a change in our personal value system.

The argument for inadequate technology is the only trivial one. The discussion in Chapters 7 and 8 clearly reveals that the technology to control pollution not only exists but, for the most part, is readily available. The other arguments are probably all valid, and, together with others that could be listed, are symbolic of the "multidisciplinary" and "holistic" approaches so often cited as necessary to understand and control pollution. Not that such a range of rationalizations for pollution constitutes an adequate answer to the question asked by the spokesman for the fifth grade class, but rather, a careful analysis of this list might help the adult world to unravel the snarl of production-population-urbanization-resource problems it has created. Perhaps an answer for the 11-year-old equal in honesty to the question asked would be that adults do not see the problem as clearly as the fifth-grader.

INEVITABILITY OF POLLUTION CONTROL

The first three chapters of this text concentrated on a description of the complex fabric of the natural environment of which man is a part. The human species alone has the unparalleled good fortune of being able to comprehend the nature of life, the forces which sustain it and the ultimate effect of its own activities on the earth's systems. The characteristics of human activity (Chapter 4) and the known effects those activities produce (Chapters 5 and 6) are an indictment of our short-sighted use of this ability. Palls of smog cover cities; streams and lakes are contaminated; and undetermined levels of metals and pesticides are telescoping their effects through world-wide food webs. As Barbara Ward has so aptly stated, "anyone can make his own priority list of evils and add the despondent conclusion that not one of them is self correcting."[1] In one sense, the control of pollution is inevitable. It is intuitively obvious that policies encouraging unlimited increases in population and dedicated to uncontrolled exploitation, development, and consumption cannot continue indefinitely without catastrophic results in a world endowed with finite resources. In the uncontrolled extreme, it would be of little importance whether the *coup de grace* were administered through famine, war, pestilence, or insidious poisoning by some as yet undiscovered toxic chemical contaminating the land, oceans, or atmosphere. Short of witless tragedy we may elect to exert positive control by changing life styles. In either event, pollution will be controlled.

The essential question is, at what point will we collectively perceive, through observation of trends and effects in the biosphere, that positive control is necessary? The presentation made in Chapter 9 established that a unanimity about the need for pollution control does not exist, even though the often tragic effects of pollution (Chapter 6) are known:

> Thousands of lives have been lost in reasonably well documented air pollution episodes (London, England; Donora, Pennsylvania).
>
> Studies reveal definite relationships between air pollution and human health; correlations between deaths from respiratory diseases and air pollution indexes have been made for parts of England, Buffalo, Staten Island, and New York City.[2] A 50 percent reduction in air pollution in the major urban centers of the United States alone would probably save $2 billion each year.[3]
>
> Financial losses from air pollution damages to crops and private property amounts to additional billions of dollars annually.
>
> Depersonalization and crowding in inner-city areas increase tensions, anxieties, and crime. Annual costs in the United States resulting from crime alone are probably over $20 million, with public expenditures for prevention and control representing additional sums.[4]
>
> Lake waters of Sweden have sustained a tenfold increase in acidity because of the sulfuric acid content of rainfall; caused by the combustion practices of central Europe.[5]

Perhaps our complacency is partly because of the fact that the estimated costs of pollution control (Chapter 8), exclusive of costs from increased crime, better housing, and welfare, are large, but not excessive. The most fanciful estimates for a national cleanup—$285 billion over 10 years—do not amount to even one half of the defense budget.[6] Of course, our complacent reaction to known pollution effects is probably also attributable to our continued belief that current policies are the best guarantee of material welfare and comfortable life styles. The range of current attitudes described in Chapter 9 indicates that the people of the United States recognize the existence of environmental threats, but for the most part do not perceive them as deadly serious.

In the absence of a clear and present danger, voluntary curtailment of production or population, or voluntary surrender of some "freedoms" to ultimately save more precious ones is unlikely. In addition, the feeble anti-pollution noises we are generating in this country must have a definite "neo-colonialist" ring in the ears of the underdeveloped nations whose only hope for higher living standards lies in industrialization and resource depletion.[6] How can we expect Third World nations to limit production and population to help solve a problem the wealthy nations created? It has even been suggested that some threshold level of material welfare (close to the U.S. average) is, in fact, prerequisite to the luxury of a desire for a clean environment.

EVALUATION OF ALTERNATIVES

Even if we are not on the brink of environmental disaster, perhaps we have learned enough from our past mistakes to become more cautious in selecting and endorsing policies affecting resource use. At least there is a vague sense of uneasiness as news releases and documentaries recite stories of oil spills, plant closures, and automobile price increases because of anti-pollution devices, smog alerts, and United Nations conferences on the environment. In his State of the Union Address in January 1971, President Nixon remarked, "Never has a nation seemed to have had more and enjoyed it less."

Caution in policy formation is clearly indicated. The basic social and political mechanisms with which alternative solutions to environmental problems can be weighed and tested have yet to be developed, although the scientific basis of goal-seeking mechanisms involving goal specification, sensing, and feedback are well established (Chapter 7). The most immediate challenge we face, then, centers on evolving effective ways to use these principles in decision-making processes. Charles Johnson has analyzed this need very accurately:

> Too many people suppose that restoring environmental quality boils down to a simple "search and destroy" mission against ecologic villains, or string up the ten most wanted polluters. What it does involve, is something far more difficult—it means initiating, in our society, an orderly system of making choices that has no precedent in all of human history.
>
> The decisions that shaped the world of the present—at least in its physical aspects—were not the result of any painful weighing of alternatives, nor did they, for the most part, involve society as a whole. They were decisions made by individuals or groups on the basis of what, in their time and place were clear-cut and valid—but limited—goals.[7]

The need to clearly specify the goals our institutional policies are designed to reach can be illustrated by several important examples. The need for energy coupled with the fact that pollution (heat, CO_2, noise) and resource depletion are directly related to energy consumption makes the energy supply business an excellent first example. Oil and gas currently account for approximately 75 percent of the energy supplies in the United States with coal (20 percent), hydropower (4 percent) and nuclear power following. In 1969, seven major oil companies, Standard Oil of New Jersey, Gulf, Shell, Texaco, Standard Oil of California, Socony Mobil and British Petroleum controlled 70 percent of the world market.[8] Small energy companies account for only 18 percent of the world market and government owned operations control the remaining 12 percent. In addition, the major oil companies are expanding into broader based energy companies—they own major coal firms, chemical, plastics, and fertilizer companies, and own a third of the world's merchant fleet. Thus, the potential for industry control exists from production through retail consumption. What sort of federal policies are currently in effect regarding the use of these valuable yet polluting natural resources? Essentially there are three:

Strict quotas on importation of cheap foreign oil. In the interest of national defense our government has encouraged the oil industry to develop supplies independent of foreign considerations.

Tax incentives which in part permit oil production companies to claim a 23-percent depletion allowance on up to 50 percent of their income. In 1967, Standard Oil of New Jersey reported a net income of $2 billion and paid taxes of only $166 million or 7.9 percent; Atlantic Richfield reported a net income of $145 million and paid no taxes.[8]

Regulations for acquiring mineral rights on public lands. The Bureau of Land Management is responsible for the management of approximately 500 million acres of public land, and under the mineral leasing act of 1920, any citizen or group of citizens may rent the annual coal rights for land in blocks of 5000 acres (up to 46,000 acres in any one state) for $10 per 5000-acre block and 25¢ an acre.

These policies are clearly superb for maximizing exploration, production and consumption of fossil fuel reserves. It is important to realize that the goals being pursued through policies administered by the Internal Revenue Service, the Bureau of Land Management and the U.S. Geological Survey are not goals of environmental protection and resource conservation.

The water supply and sewage treatment field affords a second example of the need to clearly understand the goals being served by present policy. Municipal water districts and metropolitan sewage treatment agencies (often advertised as pollution control units) pay for the costs of laying pipes and treating water or sewage by charging their customers a fee. The fee schedule is really a flat rate based upon water use, with customers at the end of miles of costly sewer pipes paying the same rate as customers closer to the treatment facility. This is an excellent example of a policy which stimulates resource use (not environmental protection) since land use in outlying areas is strongly affected by the availability of water supply and sewerage systems. Construction and development in these areas is in effect subsidized by the flat rate charge system in large metropolitan areas with centralized treatment facilities. Municipalities which overdesign present facilities, planning for growth and capitalizing on economies of scale, effectively maximize water use and encourage land development. As with the energy use example, such policies are excellent as long as it is clearly understood that the goals are not necessarily related to environmental protection. Similar arguments could be raised for the case of population control (Table 10–1).

The critical need then is to devise a means of evaluating alternative goals (Table 10–2) and the different policies required to meet them. For the preceding examples we might wish to analyze, in the interest of environmental protection and resource management, which policies would be effective for

TABLE 10–1

The Goals of Current Policies

CURRENT POLICIES	GOAL SERVED
Energy Field	
Depletion allowance, import quotas, leasing of mineral rights to public land, discount prices to large users, generation of revenue (tax) from fuel sales to construct roadways for vehicles consuming fuel.	Maximize exploration, production and consumption of fossil fuel and power generation.
Water Resources Field	
Basing of municipal sewage treatment charges on water use, flat rate sewer charges, discounting water prices for large users, low interest rates for public sewer projects.	Maximize water use and encourage land development.
Population Field	
Tax deductions for dependent children, aid to dependent children, tax incentives for marriage, repression of birth control information and services.	Maximize population growth.

limiting the use of internal combustion engines, or oil consumption, or the amount of carbon dioxide and heat discharged to the atmosphere. Which policies would be effective in providing sewer service without encouraging land development? The basic tools which these alternative policies might employ to internalize the social costs of pollution (Chapter 9) include the legal system, regulatory power, subsidies, emission charges, and establishing markets for rights of resource use.

Real and potential social costs of activities such as the operation of an automobile are effectively internalized by the use of the tort system. Difficulties appear, however, when one attempts to use the tort system to show damage to a natural resource such as the air which is, according to ancient legal opinion *res nullius*, the property of no one. Nevertheless, many of the environmental successes in the United States within the last few years have come from legal action. A most notable example of this is the Presidential stop-work order which halted the construction of the cross-Florida Barge Canal. In addition, various groups have tied up the Trans-Alaska Pipeline with litigation. Other examples are the suits against the Secretaries of the Interior and the Army to halt construction of a power plant complex at Four Corners, New Mexico; opposition to the construction of a Walt Disney Productions development—a high density recreation complex in the Mineral King area of the Sierra Nevada Mountains, and the now famous Calvert-Cliffs decision which has greatly broadened the Atomic Energy Commission's role in decisions affecting the environment.[9]

According to Joseph Sax, environmental lawyers are having some success in finding substitute concepts for "property" and "property owner."[10] Re-

TABLE 10–2

Demonstration of Policy Evaluation
Goal: To Reduce the Use of the Internal Combustion Engine

PUBLIC POLICY ALTERNATIVES	SOME LEADING CONSEQUENCES
Do nothing	Problem is likely to continue.
Tax the manufacture of the internal combustion engine $2 per horsepower	■ Would encourage the development and use of substitute engines. ■ Would also encourage low horsepower internal combustion engines which tend to be more polluting per foot-pound of work done than the more powerful internal combustion engines.
Sponsor research on technological alternatives	A technological breakthrough might produce a low-pollution or pollution-free engine that is both better and cheaper than the internal combustion engine. But then again, it might not. The research might merely produce an engine which is, say pollution free, but which has less power and is more expensive than the internal combustion engine.
Tax gasoline and diesel fuel $3 per gallon	■ Would cut back on the use of the internal combustion engine. ■ Would encourage conservation of our scarce petroleum reserves. ■ Would stimulate the search for alternate motors. ■ Would probably be violently opposed by the petroleum interests.
Tax every ton-mile of hauling done by trucks and airplanes	Would encourage shipment by rail and boat. Would probably be violently opposed by the trucking and airline industries.
Subsidize construction and operation of mass transit from the general tax till	Would make mass transit both available and cheap. If subsidy were big enough and the resulting mass transit convenient enough and cheap enough, it would tend to get large numbers of people out of their automobiles and onto public transportation, thus reducing the use of the internal combustion engine.
Divert gasoline tax and related taxes to mass transit	■ A means to subsidize mass transit with effects similar to those immediately above. ■ Would force highway construction and maintenance to compete for the general tax dollar just as mass transit has to now.
Divert gasoline tax, etc., to bicycle trails	■ A means to subsidize bicycle trails. Would make cycling much more attractive than it now is. Some people would substitute bicycles for autos some of the time. ■ Cycling is not a good alternate means of transportation for those who have poor health, or have long distances to travel, or must travel in particularly inclement weather.
Prohibit the use of the internal combustion engine in highly polluted metropolitan centers	■ Would clean up the air in that area. ■ Might create a parking problem on the fringe of that area. ■ Might create a hardship on those people who lived or had business in the prohibited area.

Reproduced with permission from Foster, Introduction to Environmental Science (Homewood, Illinois, Learning Systems Company, 1972 copyright), p. 150.

sources like the air can be defined as a commodity held in trust for the benefit of a community of citizens. Government then has the role of trustee and must implement that trust for the public good, preventing all uses which impair the interest of the beneficiaries. Tort law is one of the most rapidly changing fields of law today. The courts, expanding recognition of tort liability both in consumer-manufacturer and landlord-tenant disputes, may indicate future directions for the resolution of environmental policy conflicts. However, effective use of tort law will require much more creative responses to problems of causal relationships, injury demonstration, and damage measurement.

Regulatory power is also a partially effective policy tool, but it is costly and impractical. All automobiles, all industrial plants, or all power plants cannot be systematically inspected and analyzed for emission control without armies of inspectors. Less costly alternatives, such as less frequent inspection or relying on those regulated for compliance information, quite obviously reduce the utility of inspection. Subsidies and tax incentives for installing treatment facilities may help alleviate an immediate pollution problem, but they are less effective for controlling production than an outright charge or waste disposal fee paid by industries. In the latter case, the increased cost of production would be passed to the consumer at the market place in a more internally consistent manner. Tax incentive policies increase production costs too, but the cost is paid by higher general tax burdens and the consumer is unable to see clearly which activities are responsible for the increased costs.

The problem of determining the magnitude of specific charges for each waste discharge could be resolved by permitting the dollar value to be determined by the market mechanism. This process has long been advocated by resource economists as an effective means of allocating limited resources among competing uses. At least two factors prevent this process from being useful at this time. First, the general level of environmental degradation that would be acceptable to people of the United States and up to which pollution rights could be sold is unknown. It was emphasized in Chapter 7 that even our current levels of waste discharge are not accurately known and that a major increase in our sensing network will be required to provide that knowledge. Second, a sufficiently powerful agency must be given clear authority to own or manage the fundamental rights to resource use. As mentioned earlier, the legal system may be moving in this direction but it is not yet there.

If, for example, it was substantiated by scientific sensing and agreed to by public vote that Lake Pristine could withstand 100 pounds of BOD per day, the federal government or appropriate regional authority could auction off the annual rights to discharge up to that amount. The rights could be purchased by anyone: industry, a municipal sewage agency or a group of concerned citizens. Revenue from the rent would be used to support a strict monitoring and enforcement program for the lake and all violations would be heavily fined. New industry desiring a location on the lake would face the option of constructing appropriate treatment facilities or bidding the discharge rights away from current users. The absolute environmental impact on the lake would be controlled and the price for partial use of its waters would be whatever the market would bear. According to Foster, each producer and consumer would have to minimize his costs through the alternatives of avoiding pollution and paying no charge, or polluting to some degree and bidding in the market place for an appropriate share of the pollution rights.[11]

Standing between the current situation and the relatively enlightened process just described is a great deal of effort in monitoring waste loads and developing educational and communication techniques to encourage citizen involvement in shaping national environmental goals.

References—Chapter 10

1. Ward, Barbara.
An Urban Planet. Girard Bank, February 1971, p. 4.

2. Foster, Phillips W.
Introduction to Environmental Science. Homewood, Ill.: Learning Systems Company, Richard D. Irwin, Inc., 1972, p. 22.

3. Lave, L., and Seskin, E.
"Air Pollution and Human Health, *Science*, Vol. 169, August 1970, pp. 723–733.

4. President's Commission on Law Enforcement and Administration of Justice. *The Challenge of Crime in a Free Society.* U.S. Government Printing Office, 1967.

5. *Sweden's Case Study Contribution to the United Nations Conference on the Human Environment.* Stockholm: Royal Ministry for Foreign Affairs, 1971.

6. "The Big Cleanup." *Newsweek*, June 12, 1972.

7. Johnson, Charles.
"Law and the Environment." *Journal of Urban Law*, Vol. 48, No. 3, 1971, p. 558.

8. Ridgeway, James.
The Politics of Ecology. New York: E. P. Dutton & Co., 1970.

9. *Resources.* Resources for the Future, Inc., January 1972.

10. Sax, Joseph.
"Legal Strategies Applicable to Environmental Quality Management Decisions," *The Environmental Quality Analysis*, ed. by A. V. Kneese and B. T. Bower, Baltimore: The Johns Hopkins Press, 1972.

11. Foster, *op. cit.*, p. 131.

Index